Results and Problems in Cell Differentiation

Series Editor:
W. Henning

36

Springer-Verlag Berlin Heidelberg GmbH

D. D. Thomas · C. G. dos Remedios (Eds.)

Molecular Interactions of Actin

Actin-Myosin Interaction and Actin-Based Regulation

With 52 Figures

Springer

Dr. DAVID D. THOMAS
University of Minnesota Medical School
Department of Biochemistry
Minneapolis, MN 55455
USA

Dr. CRISTOBAL G. DOS REMEDIOS
University of Sydney
Department of Anatomy and Histology
Institute for Biomedical Research
Sydney 2006
Australia

ISSN 0080-1844

ISBN 978-3-642-08641-0 ISBN 978-3-540-46558-4
DOI 10.1007/978-3-540-46558-4

Library of Congress Cataloging-in-Publication Data

Molecular interaction of actin : actin-myosin interaction and actin-based regulation / David D. Thomas, Cristobal G. dos Remedios, eds.
p. cm. – (Results and problems in cell differentiation, ISSN 0080-1844 ; 36)
Companion volume to: Molecular interactions of actin structure and actin-binding proteins.
Includes bibliographical references and index.
ISBN 3540671110
1. Actomyosin. 2. Microfilament proteins. I. Dos Remedios, Cristobal G. II. Thomas, D. D. (David D.) III. Series.
QH607 .R4 vol. 36 [QP552.A32]

http://www.springer.de

Originally published by Springer-Verlag Berlin Heidelberg New York in 2002
Softcover reprint of the hardcover 1st edition 2002

Cover design: Meta Design, Berlin
Typesetting: Best-set Typesetter Ltd., Hong Kong
SPIN 10697338 39/3130 – 5 4 3 2 1 0 – Printed on acid-free paper

Foreword to Molecular Interactions of Actin II: Actin-Myosin Interaction and Ca Regulation of Thin Filaments

This is Volume II of a two-part book based on the Fourth Pentennial Actin Conference held at the new Sheraton Maui in Hawaii in 1997. The companion book, Volume I (Molecular Interactions of Actin: Actin Structure and Actin-Binding Proteins), was published in December 2000.

The topics covered in Volume I included the structure of actin and the regulation of cellular actin microfilaments by actin-binding proteins. Volume II focuses on the structural interactions between actin and myosin, and on Ca regulation of thin filaments in the context of muscle contraction. This Volume consists of 14 reports from researchers who attended the meeting. Unfortunately, contributions to this Volume from all the attendees were not possible, but were valuable for the success of the meeting.

We would like to thank all the authors who contributed to Volumes I and II, as well as those who attended the conference, and those who could not attend the conference but have contributed significantly to the field of actin and muscle research.

David D. Thomas and Cris dos Remedios

Contents

Actin-Myosin Interactions

An Overview of the Actin-Myosin Interactions
D.D. Thomas and O. Roopnarine

1 Structural Changes in Actin and Myosin Due to Their Strong and Weak Interactions 2
2 Fluorescence Resonance Energy Transfer in Actomyosin Complexes 2
3 Insights into Actomyosin Interactions from Actin Mutations 2
4 Role of Charges in the Actomyosin Complex 3
5 The Alanine-Scanning Mutagenesis of *Dictyostelium* Myosin II at the Ionic Interface with Actin 4
6 Familial Hypertrophic Cardiomyopathic Mutations That Affect the Actin-Myosin Interaction 4
7 Coupling Between Chemical and Mechanical Events and Conformation of Single Protein Molecules 4
References 5

Changes in Actin and Myosin Structural Dynamics Due to Their Weak and Strong Interactions
D.D. Thomas, E. Prochniewicz, and O. Roopnarine

1 Introduction 7
2 Changes in Myosin Structural Dynamics Induced by Actin 8
2.1 Global Motions of Catalytic Domain: Disorder-to-Order Transition 10
2.2 Global Motion of Light Chain Domain: Disorder-to-Order Transition, Two Angles 11
2.3 Internal Motion of the Myosin Catalytic Domain: Resolve Three Structural States 12
3 Changes in Actin Structural Dynamics Induced by Myosin 12
3.1 Global Dynamics of Actin 12
3.2 Internal Dynamics of Actin 14
4 Summary and Conclusions 15
References 17

Fluorescence Resonance Energy Transfer in Acto-Myosin Complexes
C.M. Yengo and C.L. Berger

1 Structure and Function of the Acto-Myosin Complex 21
2 Fluorescence Resonance Energy Transfer 23
3 Results and Discussion 25
References 29

Insights into Actomyosin Interactions from Actin Mutations
T.C. Doyle and E. Reisler

1 Introduction 31
2 Use of Actin Mutants in Actomyosin Studies 33
3 D24/D25 and E99/E100 in Loops 18–29 and 93–102 34
4 Acidic N-Terminus 2–5 37
5 Specificity of Actomyosin Weak Binding 41
6 Hydrophobic Strong Binding Residues 42
7 C-Terminus of Actin and the 262–274 Plug 44
8 Conclusions 45
References 47

Role of Charges in Actomyosin Interactions
P. Chaussepied and J. Van Dijk

1 Introduction 51
2 Structure of the Actomyosin Interface 51
3 Dynamics of the Actomyosin Complex 53
4 Role of the Ionic Interactions 53
5 Studies of the Ionic Interactions by Chemical Cross-Linking Experiments 54
5.1 Cross-Linking Reactions and Identification of the Cross-Linking Sites 55
5.2 Regulation of the Cross-Linking Sites by Nucleotide Analogues 56
6 A New Model for the Actomyosin Interface During the ATPase Cycle 58
7 Conclusions 60
References 61

The Alanine-Scanning Mutagenesis of *Dictyostelium* Myosin II at the Ionic Interface with Actin
H. Asukagawa and K. Sutoh

1 Introduction 65
2 Materials and Methods 67
2.1 Construction and Expression of Recombinant Myosins 67

2.2 Phenotypes of the Transformed Cells 67
2.3 Myosin Purification 67
2.4 ATPase Assays 68
2.5 In Vitro Motility Assays 68
3 Results and Discussion 68
3.1 Phenotypes of *Dictyostelium* Cells Expressing the Mutants 68
3.2 Characterization of Purified Mutant Myosins 69
References 73

Familial Hypertrophic Cardiomyopathic Myosin Mutations That Affect the Actin-Myosin Interaction
O. Roopnarine

1 Description of the FHC Disease 75
2 Myosin FHC Mutations 76
3 Myosin FHC Mutations That Affect the Actin-Myosin Interaction 77
3.1 Myosin FHC Mutations Near the Actin-Binding Interface 77
3.2 Myosin FHC Mutations Near the ATP Binding Site 80
4 Light Chain FHC Mutations That Affect the Actin-Myosin Interaction 80
5 Actin Mutations That Cause FHC 81
6 Mechanism for the FHC Disease 82
7 Conclusions 83
References 83

Coupling between Chemical and Mechanical Events and Conformation of Single Protein Molecules
Y. Ishii, A. Ishijima, and T. Yanagida

1 Introduction 87
2 Measurements of Force Exerted by Single Myosin Molecules 88
3 Visualization of Turnover of Single ATP Molecule 90
4 Coupling Between Chemical and Mechanical Events 93
5 Detection of Conformation of Single Protein Molecules 98
5.1 Fluorescence Resonance Energy Transfer of a Single Protein Molecule 98
5.2 Conformational Diversity of Myosin Subfragments 100
5.3 Dynamic Behavior of Protein Molecules 102
6 Concluding Remarks 103
References 104

Actin-Based Calcium Regulation

An Overview of Actin-Based Calcium Regulation
S.S. Lehrer

1 Introduction 107
2 Cooperativity in the Ca^{2+} Regulation of Muscle Contraction 107
3 Motility Assays of Calcium Regulation of Actin Filaments 108
4 The Ultrastructural Basis of Actin Filament Regulation 108
5 The Role of Troponin in the Ca^{2+} Regulation of Skeletal Muscle Contraction 108
6 Structural Changes Between Regulatory Proteins and Actin: A Regulation Model by Tropomyosin-Troponin Based on FRET Measurements 109

Cooperativity in the Ca^{2+} Regulation of Muscle Contraction
M.A. Geeves and S.S. Lehrer

1 Introduction 111
2 The Problem: The Force/pCa^{2+} Curve 111
3 Components of the Regulatory System 112
4 The Three Thin Filament States from Solution Studies 114
4.1 Actin-S1 ATPase: Two Activity States 114
4.2 Equilibrium Binding of S1 to Actin Filaments 115
4.3 Kinetics of S1 Binding to Actin Filaments: Three S1-Binding States 117
5 Relationship of Structural States to Solution States 120
5.1 Observation of States 120
5.1.1 Probes of Equilibrium Titrations and Kinetics 120
5.1.2 X-Ray Diffraction 121
5.1.3 Electron Microscopy 121
6 Size of the Cooperative Units 122
6.1 Cooperativity of Ca^{2+} Binding to Troponin 122
6.2 Cooperativity of Blocked to Closed Transition 123
6.3 Cooperativity of Closed to Open Transition 125
7 Relationship of Solution States to Fibers 126
References 129

Motility Assays of Calcium Regulation of Actin Filaments
M.A. LaMadrid, P.B. Chase, and A.M. Gordon

1 Introduction 133
2 Materials and Methods 135
3 Results 137
4 Discussion 139
5 Conclusion 145
References 146

The Ultrastructural Basis of Actin Filament Regulation
R. Craig and W. Lehman

1 Introduction 149
2 Regulation in Striated Muscle 150
2.1 Background 150
2.2 The "Steric Blocking" Model of Regulation 152
2.3 Early Controversies Concerning the Steric Model 153
2.4 Recent Advances on the Ultrastructural Basis of Regulation 155
2.5 Towards an Atomic Level Model for Regulation 157
2.6 Recent Criticism of the Steric Blocking Model 159
2.7 Mechanistic Insights into Regulation Using Mutant Tropomyosin 161
2.8 The Structure of Thin Filaments During Active Sliding with Myosin Filaments 161
3 Regulation in Smooth Muscle 162
3.1 Caldesmon 163
3.2 Calponin 163
4 Conclusions 164
5 Future Directions 165
References 165

The Role of Troponin in the Ca^{2+}-Regulation of Skeletal Muscle Contraction
D. Szczesna and J.D. Potter

1 The Role of TnC and Its Interaction with TnI in the Regulation of Contraction 171
2 The Role of TnT and Its Interaction with TnC in the Regulation of Contraction 176
3 Model of the Regulation of Contraction by Troponin 183
References 185

Structural Changes Between Regulatory Proteins and Actin: A Regulation Model by Tropomyosin-Troponin Based on FRET Measurements
M. Miki

1 Introduction 191
1.2 History of Ca^{2+} Regulation 191
1.2 FRET as an Optical Ruler 192
1.3 Probe Sites on Proteins 192
2 Structural Changes Measured by FRET 193
2.1 Tropomyosin–Actin 193
2.2 Troponin–Actin 196

3 A New Model For Regulation 197
3.1 FRET Data Suggest a New Regulation Model 197
3.2 Evaluation of the Regulation Model 198
4 Conclusions 200
References 201

Subject Index 205

An Overview of the Actin-Myosin Interaction

David D. Thomas and Osha Roopnarine[1]

It is proposed that force generation in muscle contraction occurs through a transition from a weakly bound state to a strongly bound state of myosin on actin during the ATPase reaction cycle. The chapters in this book probe the structural details of these interactions, with particular attention to the specific sites of interaction on each protein, the structural and dynamics properties of the bound complexes, and the differences between weak and strong interactions.

A structural model of the actin-S1 complex, based on crystal structures of the actin monomer and S1, has provided clues about the weak-binding (primarily ionic) and strong-binding (primarily hydrophobic) interactions between these two proteins (Rayment et al., 1993, Schroder et al., 1993; Milligan, 1996; Vale 1996; Mendelson and Morris, 1997). Several specific sites on myosin have been implicated in these interactions with actin: (1) the residues 626–647 of skeletal myosin (referred to as either loop 2 or as the actin-binding loop) are proposed to interact with the N-terminal residues in actin (residues 1–4) during the weak-to-strong transition of muscle contraction; (2) the helix-loop-helix, residues Gly516–His558 of the lower 50-kDa domain of myosin, is proposed to interact with actin through both ionic and hydrophobic interactions; (3) the residues Arg405–Lys415 of myosin form a flexible loop that appears to bind to actin with high affinity and are important for the actin-activated ATPase activity of myosin; and (4) the residues 567–578 (called loop 3) of myosin are proposed to be involved in hydrophobic interactions with residues 99/100 of actin.

The seven chapters in this section review recent studies on the actin-myosin interaction. To elucidate the interactions that occur in the actin-myosin interface, we must perturb the functional sites in this region of both proteins. This can be done by (1) using probes on either actin or myosin to directly measure dynamic interactions during the ATPase cycle (see Chapters by Thomas and colleagues, Yengo and Berger); (2) creating specific mutations in either actin or myosin to determine its functional effect on the actomyosin ATPase, binding, and actin motility (the chapters by Doyle and Reisler; Asugawa and Sutoh; Roopnarine); (3) capturing the actin-myosin complex by chemically

[1] Department of Biochemistry, University of Minnesota Medical School, Minneapolis, Minnesota 55455, USA

Results and Problems in Cell Differentiation, Vol. 36
D.D. Thomas, C.G. dos Remedios (Eds.): Molecular Interactions of Actin

cross-linking the complex in different chemical states using nucleotide analogs (chapter by Chaussepied and Van Dijk); and (4) simultaneously probing the mechanical and conformational changes that occur during the actomyosin interaction (see Chapter by Ishii and colleagues).

1 Structural Changes in Actin and Myosin Due to Their Strong and Weak Interactions

Conventional models suggest that the structural state of myosin is determined by the state of its bound nucleotide, which in turn determines whether it binds weakly or strongly to actin. In this first chapter, Thomas et al. ask the following questions: Do the structure and dynamics of actin depend on the binding of the myosin head and/or on the structural state (weak or strong) of the bound myosin head? Do the structure and dynamics of myosin depend on interaction with actin? This balanced approach to actin-myosin dynamics is made possible by spectroscopic analysis of site-specific spectroscopic probes attached to either actin or myosin. The results of these studies indicate that weak binding of myosin to actin produces a dynamically disordered complex that does not greatly perturb the structural dynamics of either actin or myosin, while the formation of the strong complex greatly restricts this disorder. It is likely that this disorder-to-order transition provides the driving force for actomyosin energy transduction.

2 Fluorescence Resonance Energy Transfer in Actomyosin Complexes

Yengo and Berger measured the distance between Cys374 of actin and Lys553 of S1 as a function of actin concentration using fluorescence probes. They found an increase of 5.6 Å when the S1/actin ratio increased above 0.5 under rigor conditions. These results suggest that a conformational change occurs in the C-terminus of actin due to the interaction of adjacent actin monomers that are bound to myosin S1. These results support the proposal that the myosin head interacts with two actin monomers (this concept is further discussed in the Chapter by Chaussepied and Van Dijk).

3 Insights into Actomyosin Interactions from Actin Mutations

Doyle and Reisler discuss the properties of actin mutants expressed and purified from either yeast or slime mold. This is a comprehensive review on the key residues of actin that have been implicated in either the weak- or strong-binding interaction with myosin. They review the biochemical studies of the mutated acidic actin residues (D24/D25 and E99/E100) that have been proposed to be involved in weakly bound states, which showed that the weak-binding states must be significantly populated in order to produce strongly bound states that generate force. They also summarize the results that clearly demonstrate the importance of the acidic N-terminal residues (2–5) in activating the myosin S1 ATPase. The actin residues that are involved in the hydrophobic interactions with myosin are also reviewed. Finally, the importance of the C-terminus of actin, the residues 262–274, and the segment (360–364) that is proposed to interact with the N-terminus of the alkali (or essential) light chain of myosin is discussed in this chapter.

They focused on three pairs of acidic residues on subdomain 1 of yeast actin (Asp2/Glu4, Asp24/Asp25, Glu99/Glu100) to show that they play an important role in the weak binding interactions with myosin. Using site-directed mutagenesis, they increased the charge of the N-terminus of yeast actin to be equivalent with that of rabbit skeletal actin, which resulted in increased V_{max} of the mutant actin. Their results with the N-terminal actin mutants suggest that the N-terminus may play a role in the transition of the weak to strongly bound states of myosin. The actin "charge" mutants Asp-Asn-Glu-Gln (DNEQ), ΔDSE, Asp24Ala/Asp25Ala, and Glu99Ala/Glu100Ala had lower V_{max} for actin-activated ATPase and actin motility but had the same rigor binding to S1 as the wild type actin. These results suggest that the actin mutants decreased the population of weakly bound complexes, so that fewer myosin cross-bridges attained the isomerization step with mutant actins (see Chapter by Doyle and Reisler, step 5 in Fig. 1).

4 Role of Charges in the Actomyosin Complex

Chaussepied and Van Dijk review the role of loop 2 (626–647) and 3 (567–578) of myosin during the actin-myosin interaction. They used cross-linking agents to trap the actin-myosin complex in different nucleotide states and then used chemical and proteolytic degradation to characterize the complexes. Their key result suggests that after cleavage of the β-γ phosphate bond of the bound ATP (while still in the weak-binding state), loops 2 and 3 of myosin S1 interact simultaneously and with equal affinity with two different actin monomers. This was observed regardless of the bound ADP.Pi analog (AMPPNP, ADP.BEF_x, ADP.AlF_4, or ADP.VO_4), which is surprising because AMPPNP and ADP.BEF_x have been shown to induce different lever arm orientations compared to ADP.AlF_4, and ADP.VO_4 in the crystal structures of S1. While they were unable to clearly define the sites of interaction between actin and myosin prior to ATP hydrolysis (using ATP or ATPγS), they showed that in an ATP-bound state myosin S1 interacts differently from both the weakly- (ADP.Pi) and strongly-bound states (rigor or ADP-bound). In the latter state, they observed an increase of myosin molecules interacting with a single actin monomer and a decrease of the electrostatic interactions that simultaneously involve loops 2 and 3. The significance of these results is that it suggests that the structural environment of the γPi in the active site alters the accessibility of the myosin loop structures for interaction with actin.

5 The Alanine-Scanning Mutagenesis of *Dictyostelium* Myosin II at the Ionic Interface with Actin

In their chapter, Asukagawa and Sutoh discuss the role of the 50- to 20-kDa junction by creating three myosin mutations that may affect the actin-myosin interaction. They replaced three residues with alanine (Arg620Ala, Lys622Ala, Lys623Ala) in the 50- to 20-kDa junction of myosin (loop 2 or actin-binding loop). Both the physiological myosin ATPases and V_{max} of the actin-activated ATPases were unaffected, but the actin-activated ATPases revealed that the K_m for actin binding were higher than wild type myosin. The sliding velocities of actin filaments by each mutant myosin was lower than that of wild type myosin, but increased to wild type values at higher concentrations of the mutant myosin. These results suggest that each mutation in myosin induced weaker interactions of actin-myosin in the presence of ATP. Significantly, this work suggests that the charged residues of loop 2 play an important role in actin-myosin interactions such as ATP hydrolysis, actin-myosin sliding and force generation.

6 Familial Hypertrophic Cardiomyopathic Mutations That Affect the Actin-Myosin Interaction

Familial hypertrophic cardiomyopathy is caused by single missense mutations in seven cardiac sarcomeric proteins, with the most mutations (>40) occurring in the myosin gene. The myosin mutations occur near the functional domains in the myosin head (ATP-binding site, actin-binding region, and LC-binding domain). Roopnarine reviews the functional results of several myosin mutations in its role of perturbing actin function and then correlates these results with the phenotypic manifestations of the disease in affected patients.

7 Coupling Between Chemical and Mechanical Events and Conformation of Single Protein Molecules

In their chapter, Ishii, Ishijima, and Yanagida address the question: Are the chemical states of ATPase hydrolysis directly and tightly coupled to the mechanical events that occur during muscle contraction or force displacement? They used a laser trap to measure the force exerted by myosin on an actin filament, and the single molecule imaging technique with fluorescently labeled ATP to simultaneously directly visualize the hydrolysis of ATP by myosin during the force displacement. This elegant combination of techniques allowed them to determine that the mechanical events are not always tightly coupled to chemical events, and that the energy from the ATPase reaction is stored in the actomyosin complex for later use. They also showed that the transition between metastable states is slow, which may explain the energy storage within actomyosin.

While the reconstruction of the actin-myosin S1 rigor complex has given us many clues about the actin-myosin interaction, the elucidation of the events that occur at this interface will have to be solved by biophysical and biochemical studies of the two proteins, both in solution and in the muscle fiber. This is especially important in determining the role of the highly flexible regions, such as loop 2, that are invisible to X-ray crystallography. Site-directed mutagenesis of both proteins plays a significant role in assessing the function of key residues in this region and will also allow targeted probing using various reporter groups. Spectroscopic measurements will continue to be necessary for determining dynamic events and to map the distances between residues on actin and myosin during the ATPase cycle.

References

Mendelson R, Morris EP (1997) The structure of the acto-myosin subfragment 1 complex: results of searches using data from electron microscopy and X-ray crystallography. Proc Natl Acad Sci USA 94:8533–8538

Milligan RA (1996) Protein-protein interactions in the rigor actomyosin complex. Proc Natl Acad Sci USA 93:21–22

Rayment I, Holden HM, Whittaker M, Yohn CB, Lorenz M, Holmes KC, Milligan RA (1993) Structure of the actin-myosin complex and its implications for muscle contraction. Science 261:58–65

Schroder RR, Manstein DJ, Jahn W, Holden H, Rayment I, Holmes KC, Spudich JA (1993) Three-dimensional atomic model of F-actin decorated with *Dictyostelium* myosin S1. Nature 364:171–174

Vale RH (1996) Switches, latches, and amplifiers: common themes of G proteins and molecular markers. J Cell Biol 135:291–302

them to determine that the mechanical events are not always tightly coupled to chemical events and that the energy from the ATPase reaction is stored in the actomyosin complex for later use. They also showed that the transition between metastable states is stochastic and may explain the energy storage within actomyosin.

[illegible]

[illegible]

[illegible]

Changes in Actin and Myosin Structural Dynamics Due to Their Weak and Strong Interactions

David D. Thomas, Ewa Prochniewicz, and Osha Roopnarine[1]

1 Introduction

This chapter reviews recent research on the molecular structural changes induced by the actin-myosin interaction, focusing on spectroscopic probe studies in this laboratory. We ask the following questions: What changes in myosin structure and dynamics are induced by interaction with actin? (addressed in Sect. 2), What changes in actin structure and dynamics are induced by interaction with myosin? (addressed in Sect. 3), and How are these structural changes influenced by the nucleotide-dependent strength of the actin-myosin interaction? This last is particularly important because the weak-to-strong transition (Fig. 1) generates force and movement (Eisenberg and Hill 1985). These questions are illustrated schematically in Fig. 1, which shows the actin-activated myosin ATPase scheme, emphasizing the distinction between weak and strong binding states. Although, in principle, each of the ternary complexes, A.M.N, has both weak and strong structural states (Geeves 1991; Geeves and Lehrer, this Vol.), the brackets in Fig. 1 indicate the predominant state (weak or strong) that exists in each ternary complex.

In general, we want to know what structural changes occur within the *ternary complex* A.M.N (top row of Fig. 1), using probes attached specifically to either actin or myosin, in the transition between the weak binding states (bracket on left, N = ATP, ATP-γS, ADP.P_i) and strong binding states (bracket on right, N = ADP or no nucleotide). Although this can, in principle, be observed directly by following the ATPase kinetic cycle transiently from left to right in Fig. 1, it is more common to perform an experiment that focuses on one column of Fig. 1, i.e., on a particular nucleotide (or nucleotide analog) N, mixing A with M.N and asking whether a structural change in the labeled actin or myosin is observed upon formation of the ternary complex A.M.N. The strong-binding ternary complexes are relatively easy to study, but it is difficult to isolate a complex in which most labeled actin monomers are occupied by weakly bound myosin or most labeled myosin heads are bound weakly to actin. This explains why there is much more information on strong-binding structural states than on weak-binding structural states. An alternative approach to

[1] Department of Biochemistry, Molecular Biology, and Biophysics, University of Minnesota, Minneapolis, Minnesota 55455, USA

Results and Problems in Cell Differentiation, Vol. 36
D.D. Thomas, C.G. dos Remedios (Eds.): Molecular Interactions of Actin

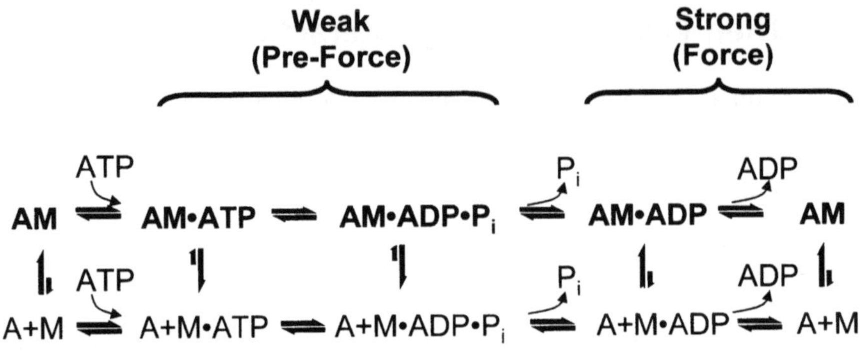

Fig. 1. Actomyosin ATPase scheme. The *horizontal* dimension shows changes in the nucleotide bound to myosin, which determines the force-generating transition from weak (*left*) to strong (*right*) states. The *vertical* dimension shows the transition between the dissociated complexes (A+M.N, *bottom*) and the bound ternary complexes (A.M.N, *top*)

studying the structural basis of actomyosin interactions is to apply perturbations (peptides or antibodies or chemical modifications) that are specific for proposed interaction sites on actin or myosin, based on structural modeling (Rayment et al. 1993b; Mendelson and Morris 1997). Although this approach is less direct than studying the actomyosin complex itself, it can provide information about the structural effects of weak interactions.

By using site-directed spectroscopic probes, we are able to probe a specific site on myosin or actin in the presence of the other protein, and in many cases, in the presence of the entire contractile apparatus. The spectroscopic probe techniques discussed here include spin label electron paramagnetic resonance (EPR), which is a high-resolution technique for the analysis of orientational and motional states, time-resolved phosphorescence anisotropy (TPA), which resolves rotational motions within large protein complexes, and fluorescence, which detects nanosecond dynamics, solvent accessibility, and distances between probes. Depending on the probe and the technique, these measurements can reveal information about: (1) global dynamics of actin monomers or myosin head domains, or (2) internal dynamics within the actin monomer or myosin head domain (Fig. 2). Global rotations of actin monomers or myosin head domains correspond to either axial (bending or tilting) motions or azimuthal (torsional or twisting) motions (Fig. 2). Some of these structural effects have been found to be cooperative, i.e., propagated along the filament so that myosin or actin molecules are affected without direct contact.

2 Changes in Myosin Structural Dynamics Induced by Actin

The myosin head is functionally divided into the catalytic domain (sometimes designated as the "motor domain", containing both the ATPase active site and the actin-binding region), and the light chain-domain (LC domain or lever

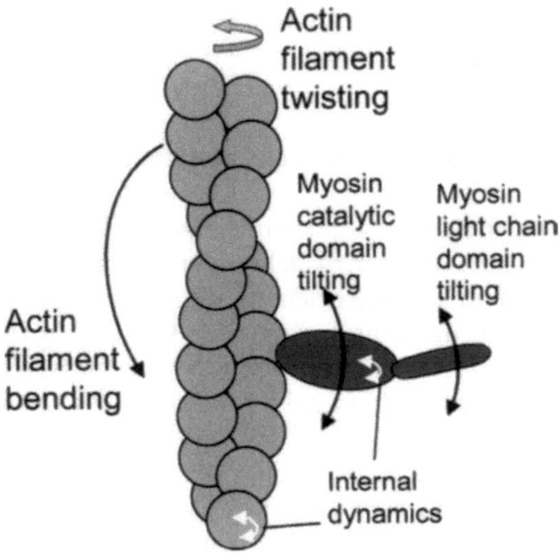

Fig. 2. Rotational dynamics of the actomyosin complex showing different modes of motion (global or internal) that are detected by spectroscopic probes. Global rotations of the actin monomer or myosin head domains correspond to either axial (bending or tilting) motions or azimuthal (torsional or twisting) motions. Internal rotations reveal information about local conformational changes that may affect other domains within the molecule

arm). The LC domain contains the essential (ELC) and regulatory light chains (RLC), which wrap around ~8 nm of an α-helix in the myosin heavy chain (Rayment et al. 1993a). By specifically attaching probes on Cys 707 (SH1) of the catalytic domain and on cysteines of the RLC, structural dynamics have been probed as a function of the states of the actomyosin ATPase cycle (Fig. 1).

In the classical model for muscle contraction, the myosin head is proposed to rotate as a rigid body from 90° to 45° during muscle contraction (Huxley 1969; Huxley and Simmons 1971). After our EPR data ruled this out (Cooke et al. 1982), models were proposed in which the catalytic domain binds to actin in a fixed rigor-like orientation and the head bends in the middle (Cooke 1986), with the LC domain acting like a lever arm (Rayment et al. 1993b). Our recent spectroscopic data have supported the model of a flexible head, and have led us to propose a model for muscle contraction in which the myosin head undergoes a disorder-to-order transition in both the catalytic domain and LC domain. Specifically, the myosin heads initially attach to actin in a weak and non-stereospecific mode that is characterized by large-scale dynamic disorder, then undergo at least two conformational transitions involving large-scale structural changes within the catalytic domain, and finally attach to actin in a strong binding and stereospecific mode to generate force. The LC domain appears to undergo a similar dynamic disorder-to-order transition (Roopnarine et al. 1998) but within a smaller orientational distribution than the catalytic domain so that the mean angle of the myosin heads is well defined (Baker et al. 1998).

2.1 Global Motions of Catalytic Domain: Disorder-to-Order Transition

We first used a maleimide spin label (MSL) attached to SH1 to show that 10–20% of the myosin heads in contracting muscle fibers are distinctly ordered with an angular orientation that is precisely the same as in rigor, the remaining heads being highly disordered as in relaxation (Cooke et al. 1982; Fajer et al. 1990a). Since there was ambiguity about whether the spin labels could directly report an axial rotation of the myosin head, we continued our search for an ideal spin label. We found that an indane-dione spin label (InVSL) bound to SH1 had its principal axis oriented almost parallel to the fiber axis so that it could directly detect any axial rotations of the myosin head (Roopnarine and Thomas 1994). Our results with this spin label confirmed that about 20% of the myosin heads in contraction are oriented as in rigor, with the remaining heads being dynamically disordered (Roopnarine and Thomas 1995). Using ATP or ATP-γ-S to trap weak-binding states revealed only dynamic disorder of the catalytic domain, with no evidence for any distinct new orientation of the myosin head either in the weak- or strong-binding states (Fajer et al. 1988, 1990b; Roopnarine and Thomas 1996). Only a very slight reorientation (5–10°), probably corresponding to a conformational change within the catalytic domain, was observed upon addition of ADP to rigor (Fajer et al. 1990b; Tanner et al 1992; Roopnarine and Thomas 1996). These results ruled out the classical model for muscle contraction in which the entire myosin head was proposed to undergo a rigid-body rotation between two large angles (Huxley 1969; Huxley and Simmons 1971). Instead, we proposed a model for contraction in which myosin heads undergo a transition from a disordered-to-ordered state (Berger and Thomas 1994; Thomas et al. 1995; Roopnarine and Thomas 1995).

Using saturation-transfer EPR (ST-EPR), we found that most of the myosin heads in contracting myofibrils or fibers were dynamically disordered in their axial orientation in the microsecond time-scale (Thomas et al. 1980; Barnett and Thomas 1989; Roopnarine and Thomas 1995). Complementary studies with time-resolved phosphorescence anisotropy (TPA) decay using phosphorescent dyes on SH1 in muscle fibers confirmed this (Stein et al. 1990). These EPR results suggested that either (1) the catalytic domain of the myosin heads had only one specific orientation of binding to actin (in the rigor-like orientation), and that the myosin heads are only dynamically disordered when detached from actin, or (2) that the myosin head has two modes of binding to actin; in a stereospecific orientation (like the rigor orientation) and another in which the heads are dynamically disordered while attached to actin. Therefore, we conducted ST-EPR experiments on spin-labeled myosin heads in solution or myofibrils under conditions where the fraction of actin-bound heads was known. We determined that actin-bound heads do undergo dynamic microsecond rotational motions during steady state ATPase activity (Svensson and Thomas 1986; Berger et al. 1989; Berger and Thomas 1993). To probe the dynamics of the weak- and strong-binding states we used nucleotide analogs (ATPγS and AMPPNP, respectively). We found that actin-bound heads are rotationally mobile in the weak-binding states, but not in the strong-binding states (Berger and Thomas 1991; Berger and Thomas 1994). This was confirmed by electron microscopy of acto-S1 during steady-state ATP hydrolysis, where it was observed that S1 attached to actin in a variety of conformations in the presence of ATP, but only in one orientation in the presence of ADP (Walker et al. 1994). EM analysis (Walker et al. 1995) and transient EPR (Thomas et al. 1995) both showed that the actin-bound S1 was disordered within 10 ms of adding ATP to the rigor complex.

2.2 Global Motion of Light Chain Domain: Disorder-to-Order Transition, Two Angles

To analyze the orientation of the LC domain, we used spin-labels bound to RLC in scallop muscle, which offers the advantage that light-chain exchange is much easier than in skeletal muscle and that the large functional effect of the RLC (conferring calcium regulation to the muscle) makes it ideal for confirming the functional integrity of labeled LC domain. We found that a spin label, bound to chicken gizzard RLC in scallop muscle, had a single orientation in rigor, but a broad orientational distribution in relaxation, with two distinct and equally populated maxima centered at 38° and 74° with respect to the muscle fiber axis. Upon activation, the same two angles were observed in contraction, but the distribution between these two angles shifted slightly, so that an additional one sixth of them were oriented as in rigor. This was the first direct observation of a distinct angle change in contraction, supporting the model that the LC domain acts as a lever arm in contraction, but also supporting the disorder-to-order transition model (Baker et al. 1998). Subsequent experiments with nucleotide analogs (Brust-Mascher et al. 1999) showed that LC domain reorientation occurs upon strong-binding to actin, but not upon the ATP hydrolysis step.

To measure the microsecond rotational motions of the LC domain, we performed ST-EPR experiments on scallop fibers containing spin-labeled *Mercenaria* RLC (labeled at Cys 50 with InVSL). We determined that the LC domain is less mobile than the catalytic domain in relaxation but more mobile in rigor, consistent with rotation of one domain relative to the other, and consistent with disorder-to-order transitions in both domains during force generation (Roopnarine et al. 1998). We further tested this by using time-resolved phosphorescence anisotropy (TPA) to measure the rotational dynamics of a phosphorescent probe on the Cys108 of gizzard RLC on scallop myofibrils. We found that rotational motions of the C-terminus of RLC are consistent with the ST-EPR results of the N-terminus, in that the LC domain is less mobile than the catalytic domain, confirming the segmental flexibility that exists between the catalytic and LC domains (Ramachandran and Thomas 1999).

2.3 Internal Motion of the Myosin Catalytic Domain: Resolve Three Structural States

Are the biochemical states of the myosin ATPase cycle coupled to specific conformational changes within the catalytic domain? We used an iodoacetamide spin label (IASL) bound to SH1 to show that conformational changes within the myosin head correspond to specific biochemical states of the ATPase cycle in solution (Barnett and Thomas 1987; Ostap et al. 1993) and in contracting muscle fibers (Ostap et al. 1995; Thomas et al. 1995). The M.ATP state (trapped by ADP.BeF$_x$ and ATPγS) showed more internal mobility than in the absence of nucleotide and less than in the transition state M.ADP.Pi (trapped by ADP.AlF$_4$ and MgADP.VO$_4$). Thus, both ATP binding and hydrolysis induce changes in the internal dynamics of the catalytic domain.

A remaining question is: What are the conformational changes that occur within the catalytic domain during its interaction with actin? As in our study of global myosin dynamics (Berger et al. 1989; Berger and Thomas 1991; Berger and Thomas 1994), we answered this question by performing EPR experiments on solutions of IASL-S1 and actin in the presence of various nucleotides, measuring the fraction of S1 bound by centrifugation, and thus calculating the spectrum of the actin-bound ternary complex A.M.N (Grinband et al. 1998). We found that actin greatly restricts the rotational mobility of IASL in strong-binding states (N = no nucleotide or ADP), producing a new conformation that is much more rigid than when myosin is free in solution. In contrast, the effect of actin on the weak-binding states of S1 was simply to perturb the distribution of states slightly, without perturbing the individual myosin conformational states.

3 Changes in Actin Structural Dynamics Induced by Myosin

3.1 Global Dynamics of Actin

The effects of myosin on the global dynamics of actin were first demonstrated by quasi-elastic light scattering (Fujime and Ishiwata 1971), but this technique could not provide site-specific information, nor could it address motions faster than the millisecond time range. Fluorescent probes and spin labels provided site-specific labeling, primarily at Cys 374, but these measurements were limited to the nanosecond time range, which was too fast. In order to obtain direct information about the microsecond dynamics of actin, it was necessary to study spin labels with saturation transfer EPR (Thomas et al. 1979; Ostap and Thomas 1991) or optical probes with time-resolved phosphorescence anisotropy, TPA (Yoshimura et al. 1984; Prochniewicz and Thomas 1999; Prochniewicz and Thomas 1997). These results showed consistently that actin filaments undergo substantial microsecond internal rotation, dominated by torsional motions, and that the strong binding of myosin heads restricts this flexibility cooperatively; i.e., maximal restriction of rotational motion at a ratio of S1 to actin as low as 0.1–0.2 (Thomas et al. 1979; Ostap and Thomas 1991; Prochniewicz and Thomas 1997). Of these studies, the only one to probe directly the difference between weakly and strongly bound states is the EPR study of Ostap and Thomas (1991). To obtain EPR data in weak-binding ternary complexes, these investigators performed EPR on a concentrated solution of S1 and spin-labeled actin (with a maleimide spin label rigidly bound to Cys 374) during the brief steady state following photolysis of caged ATP. They found that the mixture of weak-binding complexes (A.M.ATP and A.M.ADP.P_i) produced just as much restriction of the microsecond flexibility of actin as the strong-binding complexes (A.M and A.M.ADP). These results suggest that the global dynamics of actin do not depend on the weak-to-strong transition.

Phosphorescence (TPA), using erythrosin iodoacetamide attached to Cys 374, has provided detailed information about effects of myosin on both the rate and amplitude of actin's microsecond dynamics (Prochniewicz and Thomas 1997). In the absence of nucleotide (i.e., in a strong-binding state), myosin decreases the rate of actin's torsional flexibility by a factor of 4 and decreases the amplitude by nearly a factor of 2. As observed previously by ST-EPR (Thomas et al. 1979), actin's flexibility decreased nonlinearly with S1 binding, indicating that myosin's effects are propagated cooperatively beyond the monomer to which it binds. Direct analysis of weak-binding states was not attempted in this study, but weak binding was studied indirectly by using antibodies or chemical modifications specific to sites on actin that are proposed to form part of the interface responsible for weak myosin binding (Prochniewicz et al. 1993; Prochniewicz and Thomas 1997). Like strong myosin binding, these perturbations decreased actin flexibility in a cooperative manner, suggesting the propagation of effects from one actin monomer to several others. These results underscore the complex coupling that exists among the actin-myosin interface, global actin rotational dynamics, and the propagation of physical effects along the actin filament.

Other spectroscopic analyses support the propagation of global structural changes along the actin filament. Studies using both electron and optical microscopy revealed an increase in the bending flexibility of the actin filament upon binding of HMM in the presence of ATP (Takebeyashi et al. 1977; Yanagida et al. 1984), which may originate from local conformational changes in actin such as distortion or rotation of protomers and loosening of intermonomer bonds. The dissociation of acto-S1 by AMPPNP induced large undulations in the actin filament as observed using the time-resolved cryo-microscopy. This structural transition appears to originate from the release of structural constraints imposed on actin by binding of myosin heads to two successive monomers along the long-pitch helix (Menetret et al. 1991).

The development of the in vitro motility assay, which is a model of the contraction process, enabled examination of structural changes in individual actin filaments during sliding along myosin heads fixed on a glass surface. Polarized fluorescence from phalloidin-rhodamine actin during sliding indicated myosin-induced deformation of actin filaments (Borejdo and Burlacu 1994). Later fluorescence polarization measurements on sliding filaments labeled with 5-iodoacetamidotetramethylrhodamine led to the conclusion that the sliding actin filament performs an axial rotation with one revolution per 1 μm sliding distance (Sase et al. 1997).

Spectroscopic measurements on contracting glycerinated fibers were made possible by specific labeling of actin with phalloidin-FITC and measurement of polarized fluorescence. The changes in the polarization during development of isometric tension were clearly different from those accompanying transition of the fiber from relaxation to rigor and indicated a possibility that tension development is accompanied by a small axial rotation of the monomers in the actin filament (Prochniewicz et al. 1983). These results indicate that orientational changes in actin are induced by the active interaction of myosin. However, EPR experiments with spin labels attached to oriented actin using either spin-labeled phalloidin in muscle fibers (Naber et al. 1993) or a spin label attached to actin at Cys 374 in flow-oriented actin (Ostap et al. 1992) did not show a significant change in orientation upon the active interaction of myosin and ATP with actin.

3.2 Internal Dynamics of Actin

Fluorescence studies on actin labeled at the C-terminal Cys374 with IAEDANS showed that the rotational mobility of the dye was changing in a non-linear manner following addition of increasing amounts of HMM or S1 (Miki et al. 1982); this change – increase or decrease in mobility – was dependent on the kind of divalent cation (Ca or Mg) tightly bound to actin. Non-linear changes in fluorescence intensity or polarization was further observed when actin was labeled with a fluorescent nucleotide analog ε-ADP (Miki et al. 1976). These spectroscopically observed changes were interpreted as the changes in the local environment of the probe or in filament flexibility, and the non-linearity of changes indicated intermonomer cooperativity, where locally induced changes propagate along the larger segments of the filament.

The fluorescence of pyrene-iodoacetamide reacted with Cys 374 on actin has proven to be a valuable probe of actin-myosin interactions (reviewed by Geeves 1991; see Chapter by Geeves and Lehrer, this Vol.). Binding of myosin S1 in the absence of nucleotide (strong binding) decreases the intensity of pyrene fluorescence by 70%, and the change is linear with S1 binding. Although the precise physical basis of this fluorescence change is not clear, it must arise from a change in the local environment of the pyrene label near Cys 374. This label has served as a very useful probe of the kinetics of the actin-myosin interaction. In the presence of saturating ATP or ATP-γ-S (weak binding), myosin has no effect on this fluorescence signal, even under conditions where up to 70% of actin monomers have S1 bound. This result shows clearly that the structural change induced in the C terminus of actin by myosin is much greater when myosin binds strongly to actin than when it binds weakly.

There is an interesting contrast between the effects of myosin on the internal dynamics of actin (measured by pyrene fluorescence; Geeves 1991) and on the global dynamics (measured by ST-EPR; Ostap and Thomas 1991): while the internal dynamics of actin are affected only by strong myosin binding (Geeves 1991), the global dynamics are affected equally by weak and strong binding (Ostap and Thomas 1991).

Attempts to correlate changes in structural and dynamic properties of actin with the changes in its functional interaction with myosin were subsequently pursued in a comparative study on muscle actin and non-muscle actin from yeast (Prochniewicz and Thomas 1999). In this study, fluorescent probes were used to show that muscle and yeast actin have different conformational dynamics near the C terminus, and that strong myosin binding decreases these differences. If the myosin-free state of actin corresponds to the conformation of actin in the weak-binding states (left bracket in Fig. 1), then these results indicate that the transition from weak to strong (right bracket in Fig. 1) involves a larger structural change in yeast actin than in muscle actin. It was suggested that this could be the structural basis explaining why yeast actin is less effective than muscle actin in activating myosin.

4 Summary and Conclusions

Figure 3 summarizes the effects of actomyosin binding on the internal and global dynamics of either protein, as discussed in this chapter. These effects depend primarily on the strength of the interaction; which in turn depends on the state of the nucleotide at the myosin active site. When either no nucleotide or ADP is bound, the interaction is strong and the effect on each protein is maximal. When the nucleotide is ATP or ADP.P_i, or the equivalent nonhydrolyzable analogs, the interaction is weak and the effect on molecular dynamics of each protein is minimal. The weaker effects in weak-binding states are not simply the reflection of lower occupancy of binding sites – the molecular models in Fig. 3 illustrate the effects of the formation of the ternary complex, after correction for the free actin and myosin in the system. Thus EPR on myosin (Berger and Thomas 1991; Thomas et al. 1995) and pyrene fluorescence studies on actin (Geeves 1991) have shown that the formation of a ternary complex has a negligible effect on the internal dynamics of both

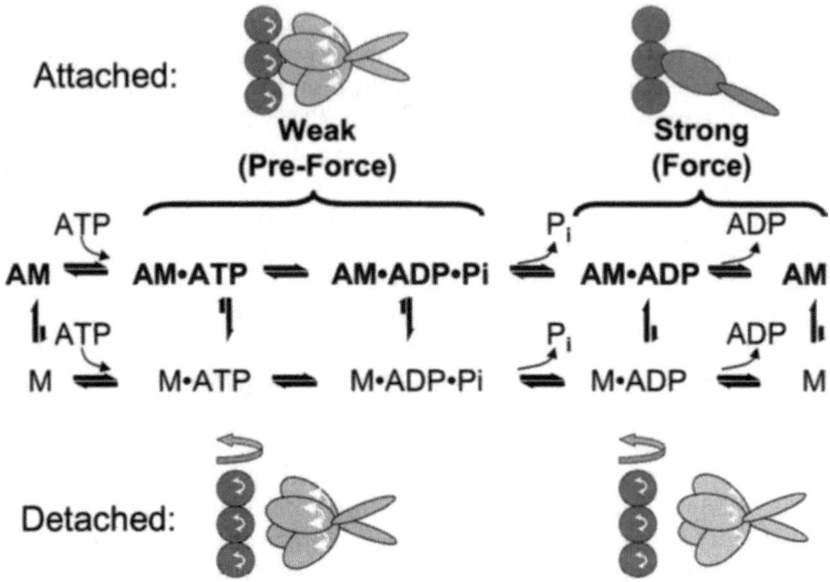

Fig. 3. Rotational dynamics of myosin heads and actin as a function of the actomyosin ATPase cycle. In the structural models, the amplitudes of *arrows* and the disorder of myosin head domains indicate the amplitudes of structural dynamics in each state. Thus the weak states (*left*) are more dynamic than the strong states (*right*), and the restriction in dynamics due to actomyosin formation is much greater in strong states than in weak

proteins (left side of Fig. 3, white arrows). As shown by both EPR (Baker et al. 1998; Roopnarine et al. 1998) and phosphorescence (Ramachandran and Thomas 1999), both domains of myosin are dynamically disordered in weak-binding states, and this is essentially unaffected by the formation of the ternary complex (left side of Fig. 3, indicated by disordered myosin domains). The only substantial effect of the formation of the weak interaction that has been reported is the EPR-detected (Ostap and Thomas 1991) restriction of the global dynamics of actin upon weak myosin binding (left column of Fig. 3, gray arrow).

The effects of strong actomyosin formation are much more dramatic. While substantial rotational dynamics, both internal and global, exist in both myosin and actin in the presence of ADP or the absence of nucleotides, spin label EPR, pyrene fluorescence, and phosphorescence all show dramatic restrictions in these motions upon formation of the strong ternary complex (right column of Fig. 3). One implication of this is that *the weak-to-strong transition is accompanied by a disorder-to-order transition in both actin and myosin,* and this is itself an excellent candidate for the structural change that produces force (Thomas et al. 1995). Another clear implication is that *the crystal structures obtained for isolated myosin and actin are not likely to be reliable representations of structures that exist in ternary complexes of these proteins* (Rayment et al. 1993a and 1993b; Dominguez et al. 1998; Houdusse et al. 1999). This is clearly true of the strong-binding states, since the spectroscopic studies indicate consistently that substantial changes occur in both proteins upon strong complex formation. For the weak complexes, the problem is not that complex formation induces large structural changes, but that *the structures themselves are dynamically disordered.* This is probably why so many different structures have been obtained for myosin S1 with nucleotides bound – each crystal is selecting one of the many different substates represented by the dynamic ensemble. Finally, there is the problem that the structures of actomyosin complexes are probably influenced strongly by their mechanical coupling to muscle protein lattice (Baker at al. 2000).

Thus, even if co-crystals of actin and myosin are obtained in the future, an accurate description of the structural changes involved in force generation will require further experiments using site-directed spectroscopic probes of both actin and myosin, in order to detect the structural dynamics of these ternary complexes under physiological conditions.

References

Baker JE, Brust-Mascher I, Ramachandran S, LaConte LEW, Thomas DD (1998) A large and distinct rotation of the myosin light chain domain occurs upon muscle contraction. Proc Natl Acad Sci USA 95:2944–2949

Baker JB, LaConte LEW, Brust-Mascher I, Thomas DD (2000) Mechanochemical coupling in muscle. Biophys J 78:2730–2739

Barnett VA, Thomas DD (1987) Resolution of conformational states of spin-labeled myosin during steady-state ATP hydrolysis. Biochemistry 26:314–323

Barnett VA, Thomas DD (1989) Microsecond rotational motion of spin-labeled myosin heads during isometric muscle contraction: saturation transfer electron paramagnetic resonance. Biophys J 56:517–523

Berger CL, Thomas DD (1991) Rotational dynamics of actin-bound intermediates in the myosin ATPase cycle. Biochemistry 30:11036–11045

Berger CL, Thomas DD (1993) Rotational dynamics of actin-bound myosin heads in active myofibrils. Biochemistry 32:3812–3821

Berger CL, Thomas DD (1994) Rotational dynamics of actin-bound intermediates of the myosin ATPase cycle in myofibrils. Biophys J 67:250–261

Berger CL, Svensson EC, Thomas DD (1989) Photolysis of caged ATP induces microsecond rotation of myosin heads on actin. Proc Natl Acad Sci USA 86:8753–8757

Borejdo J, Burlacu S (1994) Orientation of actin filaments during motion in vitro motility assay. Biophys J 66:1319–1327

Brust-Mascher I, LaConte LEW, Baker JE, Thomas DD (1999) Myosin light-chain domain rotates upon muscle activation but not ATP hydrolysis. Biochemistry 38:12607–12613

Cooke R (1986) The mechanism of muscle contraction. Crit Rev Biochem Mol Biol 21:53–118

Cooke R, Crowder MS, Thomas DD (1982) Orientation of spin-labels attached to cross-bridges in contracting muscle fibers. Nature 300:776–778

Dominguez R, Freyzon Y, Trybus KM, Cohen C (1998) Crystal structure of a vertebrate smooth muscle myosin motor domain and its complex with the essential light chain: visualization of the pre-power stroke state. Cell 94:559–571

Eisenberg E, Hill T (1985) Muscle contraction and free energy transduction in biological systems. Science 227:999–1006

Fajer PG, Fajer EA, Brunsvold NJ, Thomas DD (1988) Effects of AMPPNP on the orientation and rotational dynamics of spin-labeled myosin heads in muscle fibers. Biophys J 53:513–524

Fajer PG, Fajer EA, Thomas DD (1990a) Myosin heads have a broad orientational distribution during isometric muscle contraction. Time-resolved EPR studies using caged ATP. Proc Natl Acad Sci USA 87:5538–5542

Fajer PG, Fajer EA, Matta JJ, Thomas DD (1990b) Effect of ADP on the orientation of spin-labeled myosin heads in muscle fibers: a high resolution study with deuterated spin labels. Biochemistry 29:5865–5871

Fujime S, Ishiwata S (1971) Dynamic light scattering study of muscle F-actin. J Mol Biol 62: 251–265

Geeves MA (1991) The dynamics of actin and myosin association and the crossbridge model of muscle contraction. Biochem J 274:1–14

Grinband J, Baker JE, Thomas DD (1998) The conformational state of myosin is affected by actin in the strong binding but not in the weak binding state. Biophys J 74:A263

Houdusse A, Kalabokis VN, Himmel D, Szent-Gyorgyi AG, Cohen C (1999) Atomic structure of scallop myosin subfragment S1 complexed with MgADP: a novel conformation of the myosin head. Cell 97:459–470

Huxley HE (1969) The mechanism of muscular contraction. Science 114:1356–1366

Huxley AF, Simmons R (1971) Proposed mechanism of force generation in striated muscle. Nature 233:533–538

Mendelson R, Morris EP (1997) The structure of the acto-myosin subfragment 1 complex: results of searches using data from electron microscopy and X-ray crystallography. Proc Natl Acad Sci USA 94:8533–8538

Menetret JF, Hofmann W, Schroder RR, Rapp G, Goody RS (1991) Time-resolved cryo-electron microscopic study of the dissociation of actomyosin induced by photolysis of photolabile nucleotides. J Mol Biol 219:139–144

Miki M, Kouyama T, Mihashi K (1976) Fluorescence study of e-ADP bound to rabbit F-actin: structural change in the adenine subsite of F-actin under the influence of heavy meromyosin FEBS Lett 66:98–101

Miki M, Wahl P, Auchet JC (1982) Fluorescence anisotropy of labeled F-actin: influence of divalent cations on the interaction between F-actin and myosin heads. Biochemistry 21: 3661–3665

Naber N, Ostap EM, Thomas DD, Cooke R (1993) Orientation and rotational dynamics of spin-labeled phalloidin bound to actin in muscle fibers. Proteins 17:347–354

Ostap EM, Thomas DD (1991) Rotational dynamics of spin-labeled F-actin during activation of myosin S1 ATPase using caged ATP. Biophys J 59:1235–1241

Ostap EM, Yanagida T, Thomas DD (1992) Orientational distribution of spin-labeled actin oriented by flow. Biophys J 63:966–975

Ostap EM, White HD, Thomas DD (1993) Transient detection of spin-labeled myosin subfragment 1 conformational states during ATP hydrolysis. Biochemistry 32:6712–6720

Ostap EM, Barnett VA, Thomas DD (1995) Resolution of three structural states of spin-labeled myosin in contracting muscle. Biophys J 69:177–188

Prochniewicz E, Thomas DD (1997) Perturbations of functional interactions with myosin induce long range allosteric and cooperative structural changes in actin. Biochemistry 36: 12845–12853

Prochniewicz E, Thomas DD (1999) Differences in structural dynamics of muscle and yeast actin accompany differences in functional interactions with myosin. Biochemistry 38: 14860–14867

Prochniewicz-Nakayama E, Yanagida T, Oosawa F (1983) Studies on conformation of F-actin in muscle fibers in the relaxed state, rigor and during contraction using fluorescent phalloidin. J Cell Biol 97:1663–1667

Prochniewicz E, Katayama E, Yanagida T, Thomas DD (1993) Cooperativity in F-actin: chemical modifications of actin monomers affect the functional interactions of myosin with unmodified monomers in the same actin filament. Biophys J 65:113–123

Ramachandran S, Thomas DD (1999) Rotational dynamics of the regulatory light chain in scallop muscle detected by time-resolved phosphorescence anisotropy. Biochemistry 38:9097–9104

Rayment I, Rypniewski WR, Schmidt-Bäse K, Smith R, Tomchick DR, Benning MM, Winklemann DA, Wesenberg G, Holden HM (1993a) Three-dimensional structure of myosin subfragment-1: a molecular motor. Science 261:50–58

Rayment I, Holden HM, Whittaker M, Yohn CB, Lorenz M, Holmes KC, Milligan RA (1993b) Structure of the actin-myosin complex and its implications for muscle contraction. Science 261:58–65

Roopnarine O, Thomas DD (1994) A spin label that binds to myosin heads in muscle fibers with its principal axis parallel to the fiber axis in rigor. Biophys J 67:1634–1645

Roopnarine O, Thomas DD (1995) Orientational dynamics of indane dione spin-labeled myosin head in relaxed and contracting skeletal muscle fibers. Biophys J 68:1461–1471

Roopnarine O, Thomas DD (1996) Orientational dynamics of intermediate states of indane dione spin-labeled myosin heads in skeletal muscle fibers. Biophys J 70:2795–2806

Roopnarine O, Szent-Györgyi AG, Thomas DD (1998) Microsecond rotational dynamics of spin-labeled myosin regulatory light chain induced by relaxation and contraction of scallop muscle. Biochemistry 37:14428–14436

Sase I, Miyata H, Ishiwata S, Kinosita K Jr. (1997) Axial rotation of sliding actin filaments revealed by single-fluorophore imaging. Proc Natl Acad Sci USA 94(11):5646–5650

Stein RA, Ludescher RD, Dahlberg PS, Bennett RDL, Fajer PG, Thomas DD (1990) Time-resolved rotational dynamics of phosphorescent-labeled myosin heads in contracting muscle fibers. Biochemistry 29:10023–10031

Svensson EC, Thomas DD (1986) ATP induces microsecond rotational motions of myosin heads crosslinked to actin. Biophys J 50:999–1002

Takebayashi T, Morita Y, Oosawa F (1977) Electronmicroscopic investigation of the flexibility of F-actin. Biochim Biophys Acta 492:357–363

Tanner JW, Thomas DD, Goldman YE (1992) Transients in orientation of a fluorescent cross-bridge probe following photolysis of caged nucleotides in skeletal muscle fibres. J Mol Biol 223:185–203

Thomas DD, Seidel JC, Gergely J (1979) Rotational dynamics of F-actin in the submillisecond time range. J Mol Biol 132:257–273

Thomas DD, Ishiwata S, Seidel JC, Gergely J (1980) Sub-millisecond rotational dynamics of spin-labeled crossbridges in myofibrils. Biophys J 32:873–890

Thomas DD, Ramachandran S, Roopnarine O, Hayden DW, Ostap EM (1995) The mechanism of force generation in myosin: a disorder-to-order transition, coupled to internal structural changes. Biophys J 68:135s–141s

Walker M, White H, Belknap B, Trinick J (1994) Electron cryomicroscopy of acto-myosin-S1 during steady-state ATP hydrolysis. Biophys J 66:1563–1572

Walker M, Trinick J, White H (1995) Millisecond time resolution electron cryo-microscopy of the M-ATP transient kinetic state of the acto-myosin ATPase. Biophys J 68:87S-91S

Yanagida T, Nakase M, Nishiyama K, Oosawa F (1984) Direct observation of motion of single F-actin filaments in the presence of myosin. Nature 307:58–60

Yoshimura H, Nishio T, Mihashi K (1984) Torsional motion of eosin-labeled F-actin as detected in the time-resolved anisotropy decay of the probe on the sub-millisecond time range. J Mol Biol 179:453–467

Fluorescence Resonance Energy Transfer in Acto-Myosin Complexes

Christopher M. Yengo and Christopher L. Berger[1]

1 Structure and Function of the Acto-Myosin Complex

Myosin, the best known and most studied actin-binding protein, is a molecular motor protein that converts the chemical energy of ATP hydrolysis into mechanical work during muscle contraction, cytokinesis, and other motile functions of cells. At the molecular level muscle contraction is believed to result from the relative sliding of interdigitated thick (composed primarily of myosin) and thin (composed of actin and a host of calcium-regulatory proteins) filaments past one another. This is thought to be a cyclic process in which the hydrolysis of ATP mediates the interaction between actin and myosin (Eisenberg and Hill, 1985). The ATPase cycle of myosin consists of at least four distinct structural states. Myosin is initially bound to actin in the absence of nucleotide, or in a rigor complex (AM), at the end of the contractile cycle. Upon the addition of ATP (M.ATP), the affinity of myosin for actin is significantly reduced, causing dissociation of the acto-myosin complex. ATP is then hydrolyzed to form a second weakly bound state (M.ADP.Pi) which, like the M.ATP state, has a relatively low affinity for actin. During the process of releasing the products of ATP hydrolysis (ADP and Pi), a transition from "weak" to "strong" binding occurs, due to an increase in affinity for actin, and myosin generates force by changing the angle between the cross-bridge relative to the actin filament (i.e., the power stroke). Finally, myosin returns to the rigor state (AM), which is the highest-affinity state between actin and myosin. It is critically important to understand structurally how the binding of ATP to myosin greatly reduces its affinity for actin, and how the hydrolysis of ATP primes the cross-bridge (myosin head) for rebinding to actin in the strongly bound state and the subsequent power stroke, if chemomechanical coupling and the molecular mechanism of muscle contraction are ever to be completely understood.

The overall structure of skeletal muscle myosin II is a hexamer that is formed by two heavy chains (≈ 200 kDa) and two sets of light chains, each including a regulatory and essential light chain (≈ 20 kDa each). Myosin can be separated by proteolysis into a coiled-coil fragment of light meromyosin

[1] Department of Molecular Physiology and Biophysics, University of Vermont College of Medicine, Burlington, Vermont 05405-0068, USA

Results and Problems in Cell Differentiation, Vol. 36
D.D. Thomas, C.G. dos Remedios (Eds.): Molecular Interactions of Actin

(LMM), which forms the thick filament, and heavy meromyosin (HMM, 180 kDa), which can be separated further into two globular heads (subfragment 1 or S1) and a coiled-coil region (subfragment 2 or S2). The motile properties of myosin are contained within S1 (130 kDa) which consists of two domains; a globular region known as the catalytic domain that contains both the nucleotide binding active site and actin binding region, and a light chain binding domain (or "lever arm"), which is a unique 85-Å alpha-helix that is stabilized by the two light chains (Rayment et al. 1993a). The catalytic domain can be cleaved into three trypsin-susceptible fragments: an N-terminal 25-kDa fragment which contains the catalytic active site, a central 50-kDa fragment that is divided into upper and lower 50-kDa subdomains by a large cleft that transects the actin binding region, and a 20-kDa fragment that contains the alpha-helical light chain-binding region.

Rayment and colleagues (1993b), based on the available structural data, proposed a model of the acto-myosin complex in which three regions of skeletal muscle myosin potentially interact with actin in a sequential manner. In this model, a large flexible loop (Gly-627 to Phe-646), known as the actin binding loop (ABL), initiates binding by forming electrostatic interactions with the negatively charged N-terminus of actin. Evidence that this loop interacts with actin is that it is protected from proteolysis in the acto-S1 complex (Mornet et al. 1979), can be cross-linked to actin (Sutoh 1982), and sequence variation within the loop appears to correlate with functional differences among myosin isoforms (Uyeda et al. 1994; Rovner et al. 1995). The second interaction is thought to involve a helix-loop-helix structure in the lower 50-kDa subdomain of myosin, known as the A-site (Gly-516 to His-558), which may strengthen the acto-S1 complex by forming both hydrophobic and ionic interactions with actin. Altered actin binding and actin-activated ATPase properties of a mutant myosin containing two hydrophobic A-site residues (Met546 and Phe547) substituted with non-hydrophobic amino acids (Ser and His, respectively) have been observed (Onishi et al. 1995), and Bertrand et al. (1995) demonstrated that labeling Lys-553 in the A-site with FHS (6-[fluorescein-5(and 6)-carboxamido]hexanoic acid succinimidyl ester) could be specifically blocked when myosin was bound to actin in a rigor complex. The third and final site thought to interact with actin is a loop on the upper 50-kDa subdomain, also known as the R-site (Arg-405 to Lys-415), which may form hydrophobic interactions with the C-terminus of actin and complete the acto-myosin rigor complex. There is limited evidence that this loop interacts with actin, but one study demonstrated that synthetic peptides containing the R-site sequence bind to actin with high affinity and interfere with the actin-activated ATPase of myosin (Bartegi et al. 1997). The R-site, also known as the myopathy loop, is clinically important because a single point mutation (R403Q in beta-cardiac myosin) within the loop is one of the causes of a deadly human disease known as familial hypertrophic cardiomyopathy (FHC).

One essential feature of all models of the cyclic acto-myosin interaction is that regions of myosin that bind actin must change their conformation to alter myosin's affinity for actin during different stages of the ATPase cycle. Rayment and colleagues (1993b) proposed that the 50-kDa cleft (or actin binding cleft), which extends from the active site to the actin binding region, provides this mechanism by closing in the strongly bound states and remaining open in the weakly bound states. To examine cleft closure as well as other structural properties of myosin in different nucleotide states, X-ray crystallography on the slime mold *Dictyostelium discoideum* (Dicty) myosin II in the presence of nucleotide analogs that trap myosin in specific nucleotide states has been performed (Fisher et al. 1995; Smith and Rayment 1995; Smith and Rayment 1996). The Dicty crystal structure complexed with MgADP·BeFx was thought to mimic the pre-hydrolysis (or ATP) state, the $MgADP \cdot AlF_4^-$ and $MgADP \cdot VO_4$ structures were thought to mimic the transition state, and the skeletal S1 structure, which contained no nucleotide, was thought to represent the rigor state. Interestingly, there were no changes observed in the cleft between the skeletal S1 structure and the ADP·BeFx structure, indicating that the rigor and ATP states have similar conformations, although the ATP bound state has a low affinity for actin and the nucleotide-free state is considered a high affinity state. However, the cleft became partially closed in the $MgADP \cdot AlF_4^-$ and VO_4 structures due to rotation of the lower 50-kDa domain toward the upper 50-kDa domain. Also, the transition-state structures demonstrated a change in the reactive-thiol region, which is thought to allow rotation of the lever arm. Rayment and colleagues proposed that the cleft remains open in the nucleotide-free structure because actin may be required to close the cleft in the strongly bound state (Fisher et al. 1995). Alternatively, the recent X-ray crystallography results of smooth muscle myosin motor domain-essential light chain (MDE) in the ATP-bound ($MgADPBeF_x$) or transition-state ($MgADPAlF_4^-$) both contained a more closed actin binding cleft (Dominguez et al. 1998), consistent with the cleft being closed in the low-affinity states. Thus, either there is a distinct difference between the conformational states of different myosin isoforms, or the crystal structures may adopt more than one conformation within one nucleotide state. Thus, it is clear that alternative methods, sensitive to the structural dynamics of myosin in solution, will be required to fully understand the molecular architecture of the acto-myosin complex and the conformational changes at the acto-myosin interface responsible for the large changes in affinity between these two cytoskeletal proteins that are so critical for their function.

2 Fluorescence Resonance Energy Transfer

Fluorescence resonance energy transfer (FRET) results from the direct interaction between two dipoles, a fluorescent donor and an acceptor molecule whose absorption dipole overlaps with the emission dipole of the donor. This interaction is steeply dependent on the distance between the two dipoles ($\alpha 1/r^6$), and if the acceptor molecule is also fluorescent, energy transfer from the donor can result in the emission of fluorescence from the acceptor at donor excitation wavelengths. Thus FRET between appropriately matched donor-acceptor pairs can be used to measure intermolecular distances on the sub-nanometer scale with high precision (Wu and Stryer 1972). The efficiency (E) of fluorescence resonance energy transfer (FRET) between a donor probe and an acceptor probe can be calculated from the observed fluorescence lifetimes of the donor alone (τ_D) and in the presence of an acceptor (τ_{DA}) (Eq. 1):

$$E = 1 - (\tau_{DA}/\tau_D) \tag{1}$$

The critical Förster distance at which the efficiency of energy transfer = 50% (R_0) can be calculated from Eq. (2):

$$R_0 = [(8.79 \times 10^{23})\kappa^2 \eta^{-4} Q_D J(\lambda)]^{1/6} \tag{2}$$

where κ^2 is the orientation factor (assumed to be 2/3 for isotropic motion of the fluorescent probes during the fluorescent lifetime), η is the refractive index of the medium in which the FRET experiment takes place (assumed to be 1.4 for solutions of proteins), Q_D is the quantum yield of the donor fluorescent probe, and $J(\lambda)$ is the spectral overlap integral between the fluorescence emission of the donor probe, expressed as a fraction of the total integrated intensity [$F_D(\lambda)$], and the absorbance of the acceptor probe, expressed in units of molar extinction coefficient [$\varepsilon_A(\lambda)$; Eq. (3)]:

$$J(\lambda) = \int \varepsilon_A(\lambda) \cdot F_D(\lambda) \cdot \lambda^4 d\lambda \tag{3}$$

Thus, the distance (r) between the a donor probe and an acceptor probes can be calculated from Eq. (4) once R_0 for that particular donor-acceptor probe pair has been determined:

$$r = R_0[(1 - E)/E]^{1/6} \tag{4}$$

Previous studies with fluorescent probes on actin and myosin have attempted to measure intermolecular distances between sites on actin and myosin in a rigor complex. (reviewed by dos Remedios et al. 1987; dos Remedios and Moens 1995). The fast-reacting sulfhydryl (Cys-707 in the skeletal muscle sequence, often referred to as SH-1), has been the primary site of fluorescent labeling in myosin, and Cys-374 has been the primary site of fluorescent labeling in the C-terminus actin, for FRET studies between actin and myosin in the rigor complex. The distance measured between SH-1 in myosin and Cys-374 in actin has ranged from 4.5–6.3 nm (Takashi 1979; Trayer and Trayer 1983, 1988; Miki and Wahl 1984; Xing and Cheung 1995). However, the distance between the putative actin binding regions of myosin and actin have been more difficult to achieve, largely due to a lack of specific labeling sites in the actin-binding region of myosin. Kasprzak et al. (1989), using a synthetic peptide complementary in sequence to the 50/20-kDa junction loop, has used FRET to examine the distance from this site on myosin to Cys-374 (5.3 nm) and the nucleotide binding site (4.1 nm) of actin. FRET experiments from other regions of the putative acto-myosin interface may help to further define the molecular details of the strongly bound complex between actin and myosin.

In the current study, we have measured the distance between Lys-553 of the myosin heavy chain and Cys-374 of actin using fluorescence resonance energy transfer. Lys-553, located in the middle of the second α-helix in the putative actin-binding region of the lower 50-kDa subdomain of myosin (Rayment et al. 1993b), was labeled with the fluorescent acceptor molecule FHS as described by Bertrand et al. (1995). Actin was labeled at the C-terminus (Cys-374) with the fluorescent donor molecule IAEDANS ,5((((2-iodoacetyl)amino)ethyl) amino)-naphthalene-1-sulfonic acid, as described by dos Remedios and Cooke (1984). Distance measurements between Lys-553 of the myosin heavy chain and Cys-374 of actin in a rigor complex were made at varying molar ratios of S1 bound to actin using FRET.

3 Results and Discussion

Myosin subfragment 1 (S1) was predominantly labeled at Lys-553 of the myosin heavy chain with FHS at a molar ratio of 0.84:1 (FHS:S1, Bertrand et al. 1995), and actin was labeled at a ratio of 0.92:1 at Cys-374 with IAEDANS (dos Remedios and Cooke 1984; Crosbie et al. 1994). Modification of Lys-553 resulted in only small alterations in the actin-activated ATPase activity of FHS-S1 and no change in the ability of FHS-S1 to bind to actin in a strongly bound complex (Bertrand et al. 1995). Actin-activated ATPase and centrifuge binding experiments with IAEDANS-actin and FHS-S1 were not significantly different from the results obtained using FHS-S1 and unmodified actin described above.

The quantum yield for the IAEDANS donor probe at Cys-374 of actin was determined to be 0.30, and the overlap integral between the fluorescence emission of IAEDANS-actin and the molar absorption of FHS-S1 was determined to be $1.58 \times 10^{-13}\,cm^3M^{-1}$ (Fig. 1). Assuming that the orientation factor $\kappa^2 = 2/3$ [i.e., the donor and acceptor probes are isotropically distributed within the fluorescence lifetime of the donor IAEDANS ($\approx$ 20 ns)], R_0 was calculated to be 44 Å for the IAEDANS-actin:FHS-S1 donor-acceptor pair (Fig. 1). This is likely to be a reasonable approximation given that the anisotropy values for IAEDANS-labeled actin (Miki and Wahl 1984) and FHS-S1 (Smyczynski and Kasprzak 1997) are much smaller than the theoretical maximum of 0.40.

Results from FRET experiments performed to measure the distance between the donor IAEDANS probe at Cys-374 of actin and the acceptor FHS probe at Lys-553 of myosin S1 in a rigor complex are summarized in Table 1. IAEDANS-labeled actin (5 μM) in the absence of acceptor molecules had a fluorescent lifetime that was well fit by a single-exponential decay (19.62 ± 0.29 ns), while the addition of 0.2–1.0 mol of FHS-S1 per mol of IAEDANS-actin resulted in a biexponential decay of the donor fluorescence. The amplitude

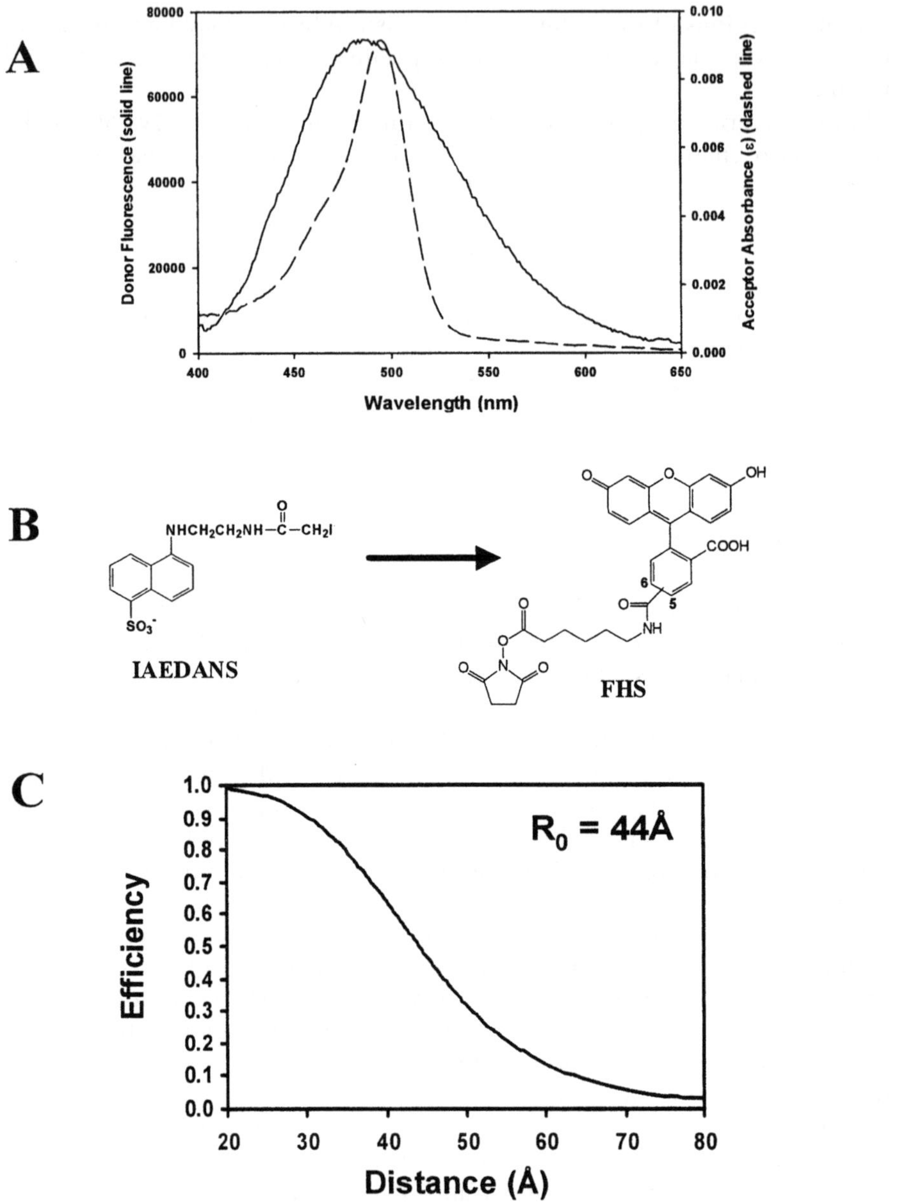

Fig. 1. **A** Fluorescence emission spectra of IAEDANS-actin (*solid line*) and molar absorbance (ε) of FHS-S1 (*dashed line*) used to calculate the overlap integral [$J(\lambda) = 1.58 \times 10^{-13}\,cm^3\,M^{-1}$; Eq. (4)]. The donor quantum yield ($Q_D = 0.30$) was also determined from the fluorescence emission spectrum of IAEDANS-actin. **B** Structures of the donor IAEDANS and acceptor FHS probes. **C** Calculated relationship between FRET efficiency (E) and the donor-acceptor distance (r) for a calculated R_0 of 44 Å [assuming $\kappa^2 = 2/3$; Eq. (4)]

Table 1. FRET (fluorescence resonance energy transfer) from IAEDANS-actin (fluorescent donor probe) in a rigor complex with FHS-S1 (fluorescent acceptor probe)

Molar ratio (S1:actin)[a]	Fluorescence lifetimes (ns)[b]		Efficiency[c]	Distance (Å)[d]
	τ_1	τ_2		
0.00	19.62 ± 0.29 (1.00)[e]	–	–	–
0.20	18.99 ± 0.32 (0.78)	1.60 ± 0.20 (0.22)	0.916 ± 0.013	29.6 ± 0.8
0.40	19.42 ± 0.44 (0.65)	1.58 ± 0.34 (0.35)	0.919 ± 0.022	29.4 ± 1.5
0.60	19.12 ± 0.28 (0.53)	4.01 ± 0.22 (0.47)	0.790 ± 0.006	35.3 ± 0.3
0.80	19.13 ± 0.34 (0.44)	3.69 ± 0.18 (0.56)	0.807 ± 0.005	34.7 ± 0.2
1.00	19.51 ± 0.29 (0.37)	3.88 ± 0.26 (0.63)	0.801 ± 0.007	34.9 ± 0.3

[a] Experiments typically contained 5 μM IAEDANS-actin and the appropriate molar concentration of FHS-S1.
[b] The fluorescence decay from IAEDANS-actin alone was fit well by a single exponential. Two exponentials were required for a good fit in the presence of FHS-S1. τ_1 was assumed to correspond to unquenched donor IAEDANS probes on actin [i.e., τ_D in Eq. (1)] and τ_2 was assumed to correspond to donor IAEDANS probes quenched by energy transfer with acceptor FHS probes on S1 [i.e., τ_{DA} in Eq. (1)].
[c] FRET efficiencies were calculated from the lifetime data according to Eq. (1).
[d] Intermolecular distances were calculated from the FRET efficiency data according to Eq. (4), using an R_0 value of 44 Å and assuming $\kappa^2 = 2/3$.
[e] All values are given as the mean ± SEM (standard error of the mean). The relative amplitudes for each lifetime component are given in parentheses (SEM for all amplitude values was ±0.03–0.04).

of the longer lifetime component decreased as an almost linear function of the amount of FHS-S1 added, with the amplitude of the shorter lifetime component increasing in a reciprocal manner. The value of the longer lifetime component was approximately 19 ns, similar to that of IAEDANS-actin alone, and the value of the shorter lifetime component depended on the molar ratio of FHS-S1:IAEDANS-actin. The fact that the relative amplitudes of both lifetime components scale approximately with the amount of acceptor added, and that the long lifetime component is fairly invariant with the addition of acceptor, suggests that FRET is predominantly occurring between the donor probe (IAEDANS) on actin and the single acceptor probe from the FHS-S1 molecule bound to that actin monomer. Thus, we interpret the longer lifetime component of our fluorescence decay to be due to unquenched donors (i.e., actin monomers without a molecule of FHS-S1 bound), and the shorter lifetime component of our fluorescence decay to be due to FRET between the donor probe on actin and an acceptor probe from the FHS-S1 molecule bound to that actin monomer.

At molar ratios less than 0.5:1 (FHS-S1:IAEDANS-actin) the shorter fluorescence lifetime component was 1.59 ± 0.03 ns and at molar ratios greater than 0.5:1 (FHS-S1:IAEDANS-actin) the shorter fluorescence lifetime component was 3.83 ± 0.27 ns. Thus, the value of the shorter lifetime component increases at higher ratios of FHS-S1 to IAEDANS-actin, suggesting an increase in the

Förster distance between the two probes from an average of 29.5 ± 1.1 Å to 35.1 ± 0.3 Å. The observed 5- to 6-Å increase in the distance between Cys-374 on actin and Lys-553 on myosin induced by increased binding of S1 to actin is most likely due to a conformational change in the C-terminus of actin that depends on the interaction with adjacent monomers in the filament that are also bound to myosin S1. Crosbie et al. (1994) demonstrated that the C-terminus of actin is quite flexible and structurally coupled to other regions within the actin filament, including the DNase I binding loop. Since S1 is believed to interact with two different monomers in the actin filament, with the primary site of interaction being the C-terminus and a secondary interaction occurring near the DNase I binding loop of actin. Thus, at molar ratios of FHS-S1 : actin greater than 0.5 : 1, it is more likely that an adjacent monomer in the actin filament will have an S1 molecule bound to it. The myosin molecule primarily bound to the adjacent actin monomer will also interact secondarily with the DNase I binding loop on the actin monomer below, altering the conformation of that actin monomer's C-terminus, including Cys-374. Consistent with our result, Moens and dos Remedios (1997) have used FRET measurements between actin monomers in an actin filament fully saturated with bound S1 molecules, and observed a similar structural change in the C-terminus of actin, a 4.5-Å change in the radial coordinates of Cys-374.

Our observed FRET distances between Cys-374 in actin and Lys-553 in myosin at any molar ratio of FHS-S1 : IAEDANS-actin are significantly longer (29–35 Å) than those calculated in models of the rigor acto-S1 complex (≈ 22 Å) constructed using the atomic coordinates of skeletal muscle myosin S1 (Rayment et al. 1993a) and actin (Kabsch et al. 1990) "docked" into electron density maps from three-dimensional cryo-EM images (Rayment et al. 1993b; Milligan 1996; Mendelson and Morris 1997). One possible explanation for the relatively long distances measured between Cys-374 in actin and Lys-553 in myosin by FRET is the unusually long linker arm on FHS (1.5 nm) that is required for site-specific labeling at Lys-553 in myosin S1 (Bertrand et al. 1995). Future studies in which other sites in the putative acto-myosin interface are fluorescently labeled for FRET studies should help to further refine the molecular architecture and structural dynamics of the strongly bound acto-myosin complex. Recent advances in site-specific labeling of the acto-myosin interface with intrinsic fluorescent probes (Yengo et al. 1998), as well as long-range luminescent energy transfer techniques (Burmeister-Getz et al. 1998), will help make this a reality.

Acknowledgements. The authors wish to thank Ivan Rayment and Robert Mendelson for making the coordinates for their atomic models of the acto-myosin complex available to us, and to the University of Vermont Muscle Club for many stimulating discussions. This work was supported in part by grants to C.L.B. from the National Institute of Health (AR-44219) and the American Heart Association.

References

Bartegi A, Roustan C, Chavanieu A, Kassab R, Fattoum A (1997) Interaction of F-actin with synthetic peptides spanning the loop region of human cardiac alpha-myosin heavy chain containing Arg403. Eur J Biochem 250:484–491

Bertrand R, Derancourt J, Kassab R (1995) Production and properties of skeletal myosin subfragment 1 selectively labeled with fluorescein at Lysine-553 proximal to strong actin binding site. Biochemistry 34:9500–9507

Burmeister-Getz E, Cooke R, Slevin PR (1998) Luminescence resonance energy transfer measurements in myosin. Biophys J 74:2451–2458

Crosbie RH, Miller C, Cheung P, Goodnight T, Muhlrad A, Reisler E (1994) Structural connectivity in actin: effect of C-terminal modifications on the properties of actin. Biophys J 67: 1957–1964

Dominguez R, Freyzon Y, Trybus KM, Cohen C (1998) Crystal structure of a vertebrate smooth muscle myosin motor domain and its complex with the essential light chain: visualization of the pre- power stroke state. Cell 94:559–571

dos Remedios CG, Cooke R (1984) Fluorescence energy transfer between probes on actin and probes on myosin. Biochim Biophys Acta 788:193–205

dos Remedios CG, Moens PDJ (1995) Actin and the actomyosin interface: a review. Biochim Biophys Acta 1228:99–124

dos Remedios CG, Miki M, Barden JA (1987) Fluorescence resonance energy transfer measurements of distances in actin and myosin. A critical evaluation. J Muscle Res Cell Motil 8:97–117

Eisenberg E, Hill TL (1985) Muscle contraction and free energy transduction in biological systems. Science 227:999–1006

Fisher AJ, Smith CA, Thoden JA, Smith R, Sutoh K, Holden HM, Rayment I (1995) X-ray structures of the myosin motor domain of *Dictyostelium discoideum* complexed with MgADP.BeFx and MgADP.AlF4-. Biochemistry 34:8960–8972

Kabsch W, Mannherz HG, Suck D, Pai EF, Holmes KC (1990) Atomic structure of the actin:DNase I complex. Nature 347:37–44

Kasprzak AA, Chaussepied P, Morales MF (1989) Location of a contact site between actin and myosin in the three-dimensional structure of the acto-S1 complex. Biochemistry 28: 9230–9238

Mendelson R, Morris EP (1997) The structure of the acto-myosin subfragment 1 complex Proc Natl Acad Sci USA 94:8533–8538

Miki M, Wahl P (1984) Fluorescence energy transfers between points in acto-subfragment-1 rigor complex. Biochim Biophys Acta 790:275–283

Milligan RA (1996) Protein-protein interactions in the rigor acto-myosin complex. Proc Natl Acad Sci USA 93:21–26

Moens PDJ, dos Remedios CG (1997) A conformational change in F-actin when myosin binds: fluorescence resonance energy transfer detects an increase in the radial coordinate of Cys-374. Biochemistry 36:7353–7360

Mornet D, Pantel P, Audemard E, Kassab R (1979) The limited tryptic cleavage of chymotryptic S-1: an approach to the characterization of the actin site in myosin heads. Biochem Biophys Res Commun 89:925–932

Onishi H, Morales MF, Katoh K, Fujiwara K (1995) The putative actin binding role of hydrophobic residues Trp546 and Phe547 in chicken gizzard heavy meromyosin. Proc Natl Acad Sci USA 92:11965–11969

Rayment I, Rypniewski WR, Schmidt-Base K, Smith R, Tomchick DR, Benning MM, Winkelmann DA, Wesenberg G, Holden HM (1993a) Three-dimensional structure of myosin subfragment-1: a molecular motor. Science 261:50–58

Rayment I, Holden HM, Whittaker M, Yohn CB, Lorenz M, Holmes KC, Milligan RA (1993b) Structure of the actin-myosin complex and its implications for muscle contraction. Science 261: 58–65

Rovner AS, Freyzon Y, Trybus KM (1995) Chimeric substitutions of the actin binding loop activate dephosphorylated but not phosphorylated smooth muscle heavy meromyosin. J Bio Chem 270:30260–30263

Smith CA, Rayment I (1995) X-ray structure of the magnesium (II)-pyrophosphate complex of the truncated head of *Dictyostelium discoideum* myosin to 2.7 Å resolution. Biochemistry 34:8973–8981

Smith CA, Rayment I (1996) X-ray structure of the magnesium(II).ADP.vanadate complex of the *Dictyostelium discoideum* myosin motor domain to 1.9 Å resolution. Biochemistry 35: 5404–5417

Smyczynski C, Kasprzak A (1997) Effect of nucleotides and actin on the orientation of the light chain-binding domain of myosin subfragment 1. Biochemistry 36:13201–13207

Sutoh K (1982) An actin binding site on the 20 K fragment of myosin subfragment 1. Biochemistry 21:4800–4804

Takashi R (1979) Fluorescence energy transfer between subfragment-1 and actin points in the rigor complex of actosubfragment 1. Biochemistry 18:5164–5169

Trayer HR, Trayer IP (1983) Fluorescence energy transfer between the myosin subfragment-1 isozymes and F-actin in the absence and presence of nucleotides. Eur J Biochem 135:47–59

Trayer HR, Trayer IP (1988) Fluorescence resonance energy transfer within the complex formed by actin and myosin subfragment 1. Comparison between weakly and strongly attached states. Biochemistry 27:5718–5727

Uyeda TQ, Ruppel KM, Spudich JA (1994) Enzymatic activities correlate with chimaeric substitutions at the actin binding face of myosin. Nature 368:567–569

Wu CW, Stryer L (1972) Proximity relationships in rhodopsin. Proc Natl Acad Sci USA 69: 1104–1108

Xing J, Cheung HC (1995) Internal movement in myosin subfragment 1 detected by fluorescence resonance energy transfer. Biochemistry 34:6475–6487

Yengo CM, Fagnant PM, Chrin L, Rovner AS, Berger CL (1998) Smooth muscle myosin mutants containing a single tryptophan reveal molecular interactions at the actin binding interface. Proc Natl Acad Sci USA 95:12944–12949

Insights into Actomyosin Interactions from Actin Mutations

Timothy C. Doyle[1] and Emil Reisler[2]

1 Introduction

One of the main goals of muscle biochemistry is to determine the molecular nature of protein-protein interactions in the force generating systems in muscle cells. This information should lead to a greater understanding of the mechanisms by which proteins produce force and motion in muscle and non-muscle cells. In the past, extensive efforts have been undertaken to clarify the interactions of actin and myosin with each other, in addition to many other proteins that interact with them, both in muscle contraction, and more generally within the actin cytoskeleton of eukaryotic cells. The actomyosin interactions are commonly analyzed in the context of two classes of complexes representing the weakly and strongly bound actomyosin states present in the cross-bridge cycle (Cooke 1997).

Force generation during the cross-bridge cycle in muscle is envisioned to depend on transitions between the weakly and strongly bound actomyosin complexes (Cooke 1997). These two states are believed to depend on ionic interactions for the weakly bound state, and on hydrophobic interactions for the strongly bound state. The transition between these two states involves changes in the myosin molecule, and maybe also in the actin filament, which alter the binding interface of these proteins. By determining the specific residues on each protein involved in the formation of the weakly and strongly bound complexes, the exact nature of these transitions should be elucidated.

Early work identified several regions within the primary structure of actin that were involved in myosin binding (dos Remedios and Moens 1995). Carbodiimide cross-linking of acto-S1 located myosin binding to the acidic N-terminus of actin and suggested the binding of the A1 light chain of S1 to residues E360, E361, E363 and E364 on actin (Sutoh 1982). Antibodies directed against peptides corresponding to short regions of the actin sequence identified two myosin head binding regions on actin, 1–7 (Miller et al. 1987) and 18–29 (Adams and Reisler 1993), and also evinced a difference between them. Such antibodies to peptides 1–7 and 18–29 were able to inhibit the weak binding of myosin and the actin-activated myosin ATPase, but had only a small

[1] Xenogen Corporation, 860 Atlantic Avenue, Alameda, California 94501, USA
[2] Department of Chemistry and Biochemistry, UCLA, 405 Hilgard Avenue, Los Angeles, California 90095, USA

Results and Problems in Cell Differentiation, Vol. 36
D.D. Thomas, C.G. dos Remedios (Eds.): Molecular Interactions of Actin

effect on the strong actomyosin binding (DasGupta and Reisler 1991, 1992; Adams and Reisler 1993). Interestingly, however, the exact nature of the inhibitions by these two antibodies seemed to differ slightly. Myosin subfragment S1 was able to bind to the same molecule of actin as the F_{ab} (Sutoh 1982; Miller et al. 1987; DasGupta and Reisler 1989; DasGupta and Reisler 1991, 1992; Adams and Reisler 1993; Dos Remedios and Moens 1995; Cooke 1997), and the presence of the antibody decreased the strong binding by about tenfold in the absence of nucleotides (DasGupta and Reisler 1992). In the presence of ADP, AMP·PNP or PP_i, the F_{ab} (Sutoh 1982; Miller et al. 1987; DasGupta and Reisler 1989, 1991, 1992; Adams and Reisler 1993; Dos Remedios and Moens 1995; Cooke 1997) caused a large decrease in the binding of S1 at low concentrations, and a cooperative increase in its binding with increasing S1 concentration (DasGupta and Reisler 1992). This implies a nucleotide-induced conformational change at the myosin binding surface of actin. Although such observations were not made with the antibody to the peptide 18–29 (Adams and Reisler 1993), this may reflect a lower affinity of the 18–29 antibody for the actin molecule ($K_{a,IgG}$ is two to three orders of magnitude smaller than $K_{a,S1}$).

A major advance in the understanding, modeling and studies of actomyosin interactions came with the X-ray crystallographic determination of the structures of first actin (Holmes et al. 1990; Kabsh et al. 1990; Lorenz et al. 1993), and then the myosin S1 fragment (Rayment et al. 1993a) and the modeling of the actomyosin complex (Rayment et al 1993b; Schröder et al. 1993). The actin structure revealed that the protein was subdivided into four subdomains, and that regions previously identified as interacting with myosin were located in subdomain 1. Indeed, within the models of actin filaments that have been proposed, these same regions are located on the exposed surface of the filament. However, it was the subsequent X-ray crystallographic determination of the structure of S1, and the modeling of the myosin heads on the actin filaments (Rayment et al 1993b; Schröder et al. 1993), at least in the rigor state, which have provided the current framework for analyzing the structure of acto-S1 complexes (see Sheterline and Sparrow 1994 for review). The models implicated the N-terminus region of actin and the amino acids 18–29 identified in antibody studies as playing important roles in direct actomyosin contacts. The models suggested also that actomyosin binding appeared to be stabilized by electrostatic interactions between the lysine-rich loop 626–647 of myosin and the N-terminus of actin, as well as the acidic pair of residues D24/D25 in loop 18–29 of actin. In addition, it was suggested that a second electrostatic site on actin at the C-terminal end of helix 79–92, and/or E99/E100 in loop 93–102 and E332/R333/K334 interact with myosin loop 403–415. Myosin was proposed also to bind to residues 38–42 on the DNase I loop of actin (Rayment et al. 1993b), but this suggestion was contested on the basis of antibody work and has not yet been tested in mutational studies (Kim et al. 1996). A stereospecific hydrophobic site involving residues A144, I341, I345, L349 and F352 of actin was proposed to bind to hydrophobic residues on myosin 529–543. All the actin residues identified were thus located on the same surface, or very close to, subdomain 1 (with the exception of the H40 and G42 on the 38–52 loop). However, in considering these assignments on actin, it should be remembered that the structure of myosin did not reveal clearly the exact location of the side chains of the amino acids, and certain regions of the myosin molecule implicated in actin binding were not visualized in the structure. Moreover, changes in myosin structure associated with actin binding have yet to be elucidated. Thus, unambiguous assignment of residues important in actomyosin interactions was not possible. Another limitation of the structural model of acto-S1 was that it was proposed for the rigor binding state, and many of the charged residues implicated in the interaction were postulated to take place in the weakly bound states. Of course, the rigor binding state does not preclude the involvement of the same ionic interactions, but further experiments to investigate the specific residues involved in both the weakly and strongly bound states were needed. Such experiments, comprising detailed mapping of actin residues at the actomyosin interface became feasible with the expression of actin and myosin mutants.

2 Use of Actin Mutants in Actomyosin Studies

The use of molecular biology techniques to introduce specific mutations in the actin sequence has aided the mapping of residues involved in binding myosin, as well as a number of other actin binding proteins both in muscle and cytoskeleton. Skeletal α-actin has not been the target of this mutational analysis, due to difficulties in performing such experiments in higher organisms. Although chicken β-actin has been successfully heterologously expressed in the yeast *Saccharomyces cerevisiae* (Karlsson et al. 1991), this protein confers detrimental phenotypes onto the strain, which might be further exacerbated, possibly to lethality to the cell, by the introduction of additional mutations to the gene. The chicken gene, however, is capable of viable cell growth in the absence of the native yeast protein. A similar expression of the skeletal α-actin has not been reported.

Attention, therefore, turned to the use of actins from lower eukaryotic cells, such as the slime mold *Dictyostelium discoideum* and the brewer's yeast *Saccharomyces cerevisiae*. The validity of using these actins in investigating events of muscle contraction has been well documented. Both organisms contain endogenous myosins, which although not necessarily involved in cellular contraction, are involved in movement of cellular organelles within the organism. Actin is also one of the most highly conserved proteins in eukaryotic cells, with an 88% sequence identity of the yeast actin to mammalian muscle and cytoplasmic actins (Gallwitz and Suers 1980; Ng and Abelson 1980). Most convincingly, however, yeast actin filaments are capable of movement on glass cover slips coated with rabbit muscle myosin in the presence of ATP (Kron et al. 1992). Yeast also has the advantage of only containing a single, essential actin gene, and with the easy genetic manipulation of this organism, a large number of targeted mutations may be made and purified without contamination by other actin isoforms.

Although several groups have made specific mutations in the yeast actin gene to investigate their effects on actomyosin interactions, a large collection of mutants were made to act as a more general resource for investigations of actin function in the yeast cytoskeleton and beyond this. Most notably, a clustered charged-to-alanine scanning mutagenesis (Cunningham and Wells 1989) was performed on the actin gene (Wertman et al. 1992), targeting charged residues in the primary sequence and mutating them to the small, uncharged alanine. It was argued that these mutations would probably be on solvent-exposed regions and would be relatively inconsequential to the protein structure. For actin, along with other proteins mutated by this strategy (e.g., Bass et al. 1991; Bennett at al. 1991), most of the mutations indeed turned out to be solvent exposed when the structure was subsequently published, and the large number (23 of 35 made) of mutants producing viable proteins confirmed that the mutations did not alter the structure of the protein dramatically. Along with other mutations made by other groups (e.g., Johannes and Gallwitz 1991; Cook et al. 1992, 1993; Miller et al. 1996a), all of the charged residues implicated in actomyosin binding had been targeted for substitution, allowing biochemical analysis of most of these regions. However, one region, that of loop 332–334, produced an inviable protein (Wertman et al 1992), so further investigation of this potentially very interesting region of the protein has not been undertaken to date. Charged residues D24/D25 and E99/E100, as well as those at the N-terminus of actin have been extensively investigated by several laboratories, using both yeast (Cook et al. 1992, 1993; Miller and Reisler 1995; Miller et al. 1996b) and *Dictyostelium* (Sutoh et al. 1991; Johara et al. 1993) proteins. In addition, new mutations at residues implicated in strong, hydrophobic interactions have also recently been constructed and their myosin-linked function has been examined (Miller et al. 1996a). This review will concentrate on each of these sites in turn (described in Table 1 and shown in Fig. 1).

3 D24/D25 and E99/E100 in Loops 18–29 and 93–102

The two pairs of acidic residues, D24/D25 and E99/E100, and their parent loops 18–29 and 93–102, show an almost total sequence conservation across many species (see Sheterline and Sparrow 1994), and would therefore appear to play important roles in protein-protein interactions. Initial investigations into the role of charged loop residues were performed by Johara et al. using the *Dictyostelium* actin 15 protein (Johara et al. 1993), in which the acidic residues were changed to histidines. Four regions of the sequence were targeted in this study, D24/D25, E99/E100, E360/E361 and E363/E364, all of which lie close together on subdomain 1. Although no specific phenotype was reported in the cells expressing these mutants, it should be remembered that these cells

Table 1. Sequences of the main actin mutants discussed in this review. Mutations of interest are shown in bold and italicized

Actin	N-terminus	24/25	99/100	341/345
Rabbit skeletal	Ac-D-E-D-E-T-	-D-D-	-E-E-	-I- -I-
Dictyostelium discoideum Act15p				
Wild type	Ac-D-G-E-D-V-	-D-D-	-E-E-	-I- -I-
D24H/D25H	Ac-D-G-E-D-V-	***-H-H-***	-E-E-	-I- -I-
E99H/E100H	Ac-D-G-E-D-V-	-D-D-	***-H-H-***	-I- -I-
D1H	Ac-***H***-G-E-D-V-	-D-D-	-E-E-	-I- -I-
D1H/D4H	Ac-***H***-G-E-***H***-V-	-D-D-	-E-E-	-I- -I-
Saccharomyces cerevisiae Act1				
Wild type	Ac-M-D-S-E-V-	-D-D-	-E-E-	-I- -I-
D24A/D25A	Ac-M-D-S-E-V-	***-A-A-***	-E-E-	-I- -I-
E99A/E100A	Ac-M-D-S-E-V-	-D-D-	***-A-A-***	-I- -I-
DNEQ	Ac-M-***N***-S-***Q***-V-	-D-D-	-E-E-	-I- -I-
ΔDSE	+$\boldsymbol{H_3N}$-V-	-D-D-	-E-E-	-I- -I-
4Ac	Ac-M-D-E-D-E-V-	-D-D-	-E-E-	-I- -I-
4Ac/D24A/D25A	Ac-M-D-***E-D***-E-V-	***-A-A-***	-E-E-	-I- -I-
4Ac/E99A/E100A	Ac-M-D-***E-D***-E-V-	-D-D-	***-A-A-***	-I- -I-
I341A	Ac-M-D-S-E-V-	-D-D-	-E-E-	***-A-*** -I-
I345A	Ac-M-D-S-E-V-	-D-D-	-E-E-	-I- ***-A-***

express a large number of different actins. This required the separation of mutants from actin isoforms during purification to allow biochemical characterization of the mutant proteins. The analysis of Johara et al. (1993) showed that the C-terminal mutations only slightly affected myosin binding, actin-activated S1-ATPase, motility and the force generated by actin filaments. However, the 24/25 mutant protein showed no activation of S1-ATPase nor motility, and the 99/100 mutation also caused a large decrease in motility, and the activation of S1-ATPase.

In a more extensive study, the effect of several charge-to-alanine substitutions in yeast actin on actomyosin interactions was examined (Miller et al. 1996b; Miller and Reisler 1995). The mutants in helix 79–92 (D80A/D81A and E83A/K84A) showed similar properties to the wild-type actin for all actomyosin parameters determined. The D24A/D25A and E99A/E100A loop mutants showed decreased S1 ATPase activation and lack of motility in the in vitro assays (Miller and Reisler 1995; Table 2), but the addition of the viscosity-enhancing agent methyl cellulose to these assays restored the mutant motility to that of wild-type yeast actin. The motion of these mutants, but not that of the wild type, was also stopped at high ionic strength conditions (μ = 150 mM) which do not favor weak actomyosin interactions (Miller et al. 1996b). Although the filaments of these mutants bound to HMM-coated coverslips in the absence of ATP in the motility assays, addition of ATP caused the loop mutants to diffuse away from the HMM, either in the absence of methyl

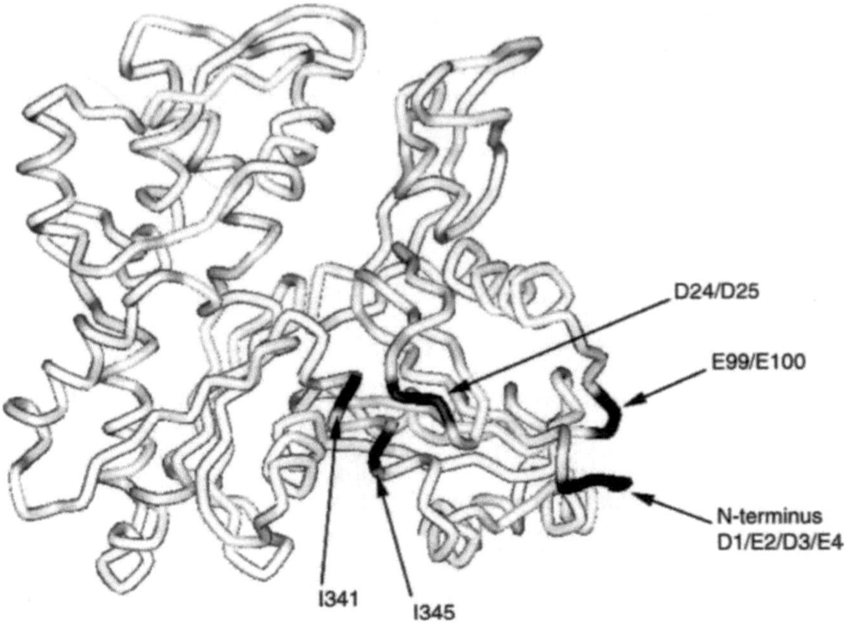

Fig. 1. Location of mutated amino acid residues discussed in this review on the G-actin ribbon structure (Kabsh et al. 1990; Lorenz et al. 1993). Residues implicated in binding myosin, and subjected to mutagenesis, are shown in *black*. Figure was generated using MSI WebLab ViewerLite 3.1

Table 2. Relative speeds of loop mutants and wild-type actins in the in vitro motility assays. The speeds of mutant actin filaments are normalized to that of wild-type actin in the respective studies. (Johara et al. 1993; Miller et al. 1995)

Actin	–MC[a]	+MC	+MC (150 mM salt)
WT	100	100	100
360H/361H	130	ND	ND
363H/364H	121	ND	ND
99H/100H	17[b]	ND	ND
24H/25H	0[c]	ND	ND
80A/81A	103	103	ND
82A/83A	106	97	ND
99A/100A	0[c]	100	0[c]
24A/25A	0[b,c]	105[b]	0[b,c]

[a] *MC*, addition of 0.7% methyl cellulose to assays.
[b] Actin filaments bundled.
[c] Filaments diffused away from surface.

cellulose (at μ = 50 mM) or under high salt conditions (μ = 150 mM). Even in the presence of methyl cellulose, the tails of moving filaments of these mutants showed a tendency to "wave back and forth", suggesting impaired binding. These results implied that although rigor binding was unaffected by the D24A/D25A and E99A/E100A mutations, weak binding was compromised significantly. In these assays methyl cellulose is envisioned to allow the filaments to remain in the vicinity of the HMM long enough for the passage through the cross-bridge cycle, before diffusion removes the filaments from the HMM-coated surface. Thus, it is apparent that substitution of the D24/D25 or E99/E100 with either alanines or histidines is sufficient under some conditions to affect the motility of actin filaments, emphasizing the importance of these charges in weak actomyosin interactions.

Binding studies confirmed these observations. Rigor binding of S1 to the actin mutants was identical, whilst weak binding of the D24A/D25A and E99A/E100A mutants was decreased between two- and threefold, and that of the 99/100 mutation by half (Miller and Reisler 1995). In analogy to the histidine mutants, the actin-activated S1-ATPases of the D24A/D25A and E99A/E100A mutants were also greatly decreased, whilst those of the helix mutations (D80A/D81A and E83A/K84A) were unaffected. The low quantities of ATPases prevented accurate K_m and V_{max} determination for the loop mutants.

Both pairs of mutations clearly implicate the 24/25 and 99/100 regions as important in the weak actomyosin interaction, but with minimal contribution to the rigor state. Changes in the in vitro motility assay conditions which decrease weak actomyosin interactions (such as removal of a viscosity agent, or an increase in salt concentration) alter the binding ability and motile function of these mutants. However, no significant difference between the two loop mutants was found, at least with the alanine substitutions. This might suggest that each loop contributes equally to the weak binding to myosin, in concert with the acidic charges at the N-terminus, which are discussed next.

4 Acidic N-Terminus 2–5

One of the most significant differences between actin isoforms, both within an organism and across species, is the number of acidic residues present at the N-terminus of the protein. Skeletal muscle α-actin has four acidic residues (1D, 2E, 3D, 4E), whilst α-actin and *Dictyostelium* actins have three, and the yeast protein has just two (see sequence comparisons in Sheterline and Sparrow 1994). Skeletal actin shows increased actin-activated S1 ATPase, compared with the yeast protein, which may, in part, be a consequence of these extra charged residues (Cook et al. 1993). Antibody studies using $F_{ab(1-7)}$ described earlier indicated that this region is involved in ATP-dependent weak binding, and has a small effect on the rigor binding of S1 (DasGupta and Reisler 1989). Several groups have used mutations at the N-terminus of actins to investigate the role of these residues in the weak and rigor actomyosin interactions.

The first studies to show that mutations to the N-terminus of actin affected S1 binding used derivatives of α-actin expressed in yeast cells (Aspenstrom and Karlsson 1991). This protein has three aspartate residues at the N-terminus, and the effects of either deleting the second and third residue, or substituting these residues with lysine and alanine residues, were investigated. All these proteins showed normal polymerization and binding to DNase I, but had altered ability to bind S1 heads as determined by electron microscopy observations. Deletion or alanine substitution mutants showed fewer S1 heads compared with control actin, whilst the lysine substitution appeared to abolish S1 binding totally. However, no further characterization of these proteins has been reported. Mutations of the yeast acidic N-terminal residues to valine (D2V, E4V and D2V/E4V) have also been generated (Johannes and Gallwitz 1991), and these produce viable proteins, but no biochemical analysis has been performed on them.

Sutoh et al. (1991)generated mutations of the *Dictyostelium* actin N-terminus acidic residues in a similar fashion to the procedure which they employed for substitutions of the 24/25 and 99/100 acidic residues in the loops. The act15 gene has three acidic residues at the N-terminus, and either the first, or first and third of these were mutated to histidine residues (wt: D-G-E-D, D1H: H-G-E-D, D1H/D4H: H-G-E-H*). These mutants showed increasing loss of actin-activated ATPase with increasing histidine substitutions (from 14.6 to 1.2 s^{-1}), although the affinities of the different actins for myosin S1 did not appear to be altered, both by kinetic and binding studies (Sutoh et al. 1991). In analogy to decreasing ATPase activities, motilities of the different actins were also decreased, with the double mutant showing the greatest effect (~80% decrease). Accordingly, Sutoh et al. (1991) concluded that the ionic interactions between the N-terminus of actin and myosin are important for the ATP-dependent actomyosin interaction, in some way activating the myosin ATPase within the cross-bridge cycle.

Rubenstein's group has generated a number of N-terminal mutations to yeast actin, removing the charge residues either by substitution (DNEQ for D2 N/E4Q) or deletion (ΔDSE; Cook et al. 1992), as well as introducing the rabbit skeletal N-terminal sequence (4Ac for S3[ED], Cook et al. 1993). Interestingly, the DNEQ mutant protein showed an in vitro bundling phenotype similar to that seen for *Dictyostelium* E99H/E100H and yeast D24A/D25A mutants, and both the DNEQ and ΔDSE mutants showed a secretion defect in yeast cells, suggesting a myosin-related yeast phenotype (Cook et al. 1992). N-terminal analysis of the ΔDSE mutant showed that the initiator methionine had been removed, unlike other yeast actins, and that N-terminal acetylation had not been performed in this strain, resulting in a positive charge to the N-terminus

* For most actins, N-terminal processing of actin acetylates and then removes the initiator methionine residue, and then acetylates the next residue. If the second residue is a cysteine (class 2 actins), this is similarly processed. However, most yeast species do not remove the first residue, although they will acetylate the methionines (Cook et al. 1991).

of this protein. DNEQ, however, showed the usual acetylation and retention of the N-terminal methionine (Cook et al. 1992).

Initial investigation of the actin-activated S1-ATPase of the DNEQ and ΔDSE actins confirmed the decrease observed with the acidic-to-histidine mutations of Sutoh et al. (1991), the DNEQ and ΔDSE mutants showing less than threefold activation, compared with a 15.3-fold activation at the same concentration of the wild-type yeast protein (Cook et al. 1992). However, direct comparison of the results of these different studies is hampered by the exact nature of the mutations introduced. Whilst the yeast mutations neutralize the N-terminus either by a conservative substitution, or deletion of the charged residues, the *Dictyostelium* mutants retain at least one acidic residue, and substituted residues are replaced with a bulky histidine residue, which may affect the structure of the N-terminal region of the protein. In addition, as discussed below, the extra acidic residue present in the N-terminus of the *Dictyostelium* actin may itself improve the ATPase activation, compared with that achieved by the yeast protein. Despite these qualifications, the reasons for the low activation of myosin ATPase by the N-terminal actin mutants became clearer after consideration of the results obtained on the actins cross-linked to S1 by dimethyl suberimidate.

The activation of ATPase in the dimethyl suberimidate cross-linked samples of S1 and DNEQ, ΔDSE and wild-type actins paralleled reaction time (and hence degree of cross-linking), and showed similarly large increase in activity for all actins. Thus, the mutant actins show a decreased activation largely due to a decreased affinity for S1 in the presence of ATP, leading to the conclusion that the primary role for the two acidic residues in the N-terminus is to contribute to the weak binding of actin to S1. However, since the activities of the DNEQ and ΔDSE mutants remain reduced even after cross-linking, these proteins are probably also catalytically deficient, with V_{max} values for S1 ATPase approximately half that of the wild-type actin.

Addition of extra acidic residues to the N-terminus of yeast actin, generating a sequence similar to the rabbit skeletal protein (termed 4Ac) showed that these extra charges do indeed contribute to the increased activation of the S1-ATPase observed for the skeletal protein compared with the native yeast protein (Cook et al. 1993). ATPase activations by actin were determined under standard ([KCl] = 25 mM, K_m too high to determine) and low salt conditions ([KCl] = 5 mM, allowed determination of K_m and V_{max}), and in both cases the 4Ac mutant yielded a threefold increased S1-ATPase activation, and the rabbit actin a ninefold increase compared with the wild-type yeast protein. Clearly, the additional pair of acidic residues enhances the activation of the S1-ATPase in the 4Ac mutant, but other determinants on the rabbit actin must further optimize its activation of S1 ATPase.

All these results, with both *Dictyostelium* and yeast mutants, clearly demonstrate that S1-ATPase activation by actin is linked to the charge at the N-terminus of actin. It has been proposed that the binding of the acidic residues by the myosin lysine-rich 626–647 loop confers a conformational change in the myosin head (Holmes 1995; Miller et al. 1995), facilitating the activation of the ATPase. If this is true, it is easy to rationalize that increasing the charge at the N-terminus of actin may aid and optimize this movement.

The yeast N-terminal mutant actins also showed altered motilities in the in vitro assays. Whilst wild-type yeast, 4Ac mutant and rabbit skeletal actins moved at essentially similar speeds, the DNEQ and ΔDSE mutant actins had similar properties to the yeast D24A/D25A and E99A/E100A mutants (Miller et al. 1996b). In the absence of the viscosity enhancing agent methyl cellulose, the mutant proteins diffuse away from the myosin heads upon addition of ATP to the assays, but in its presence, and at low ionic strength (μ = 50 mM), the speeds of all the actins are similar. Increasing the salt concentration to more physiological levels (μ = 150 mM), leaves only the wild-type and 4Ac yeast actins attached and moving over HMM at normal speeds upon addition of ATP, whilst the other mutants diffuse away from the surface. These results are somewhat in conflict with the *Dictyostelium* N-terminal histidine substitutions (Sutoh et al. 1991), which remain motile in the absence of methyl cellulose, although the velocity of the filaments decreases as sequential substitutions are made. As discussed above, this may be due to the presence of at least one acidic residue in these mutants, which may allow for some weak binding under the motility assay conditions.

Eluded to above is the effect of the N-terminus mutations on S1 binding. A significant observation was that of the effect of the DNEQ and ΔDSE mutations on both weak and rigor actin binding to myosin. In the presence of ATP, binding of both the DNEQ and ΔDSE mutants is decreased by about threefold (Table 3), similar to the decrease in weak binding observed with the loop mutants D24A/D25A and E99A/E100A. This, not surprisingly, also implicates the two acidic residues at the N-terminus of actin in weak binding to myosin. In the absence of nucleotides, both the DNEQ and ΔDSE actins also show a three- to fourfold decrease in rigor binding (Table 3). This is unlike the results obtained for the loop mutants, which show unaltered rigor binding, and suggests that whilst the loop acidic residues have a role only in the weak binding stage, the N-terminal acidic residues contribute to myosin binding throughout

Table 3. Binding of S1 to N-terminal yeast actin mutants. K_a values are taken from Miller et al. (1996b)

	Nucleotide	[KCl]	K_a
Wild type	–	100 mM	$1.8 \pm 0.2 \times 10^6\,M^{-1}$
DNEQ	–	100 mM	$0.7 \pm 0.5 \times 10^6\,M^{-1}$
ΔDSE	–	100 mM	$0.4 \pm 0.5 \times 10^6\,M^{-1}$
4Ac	–	100 mM	$1.7 \pm 0.2 \times 10^6\,M^{-1}$
Wild type	3 mM ATP	5 mM	$4.6 \pm 0.1 \times 10^4\,M^{-1}$
DNEQ	3 mM ATP	5 mM	$1.6 \pm 0.1 \times 10^4\,M^{-1}$
ΔDSE	3 mM ATP	5 mM	$1.6 \pm 0.1 \times 10^4\,M^{-1}$

the cross-bridge cycle. The 4Ac mutant actin, however, showed no change in binding in either the rigor or weak states of the cross-bridge cycle. This is not surprising considering the fact that wild-type yeast and rabbit skeletal proteins have similar binding affinities for S1. Addition of two extra charges to the wild-type yeast actin N-terminus does not increase the weak binding in the 4Ac mutant, but rather improves the activation of the S1-ATPase. The implication of this result is that the third and fourth acidic residues at the N-terminus of yeast actin facilitate the isomerization of the weakly bound acto-S1 complex to a more strongly bound state, i.e., they participate in and influence the transition between the weakly and strongly bound actomyosin states.

Preliminary insight into mechanical consequences of these differences in binding affinities of actin for S1 was obtained from the in vitro motility assays, carried out either in the presence of external load i.e., modified HMM, or decreasing HMM concentrations, such that movement of actin filaments stops at a point at which the force generated by cross-bridges is insufficient to sustain motion and/or overcome the intrinsic load of the modified HMM (Miller et al. 1996b). These experiments provide information on relative forces generated by actin mutants with myosin. Although the 4Ac mutant showed greater force production than wild-type actin in these assays, the DNEQ and ΔDSE mutants showed no difference from the loop mutants, and generated the same force as each other, but less than the wild-type actin. Thus, although the N-terminal acidic residues play a role in binding throughout the cell cycle, their relative contribution to rigor binding is too small to affect force generation.

In assays performed on all of the different yeast actins, the charged pairs D24/D25 and E99/E100 are functionally indistinguishable. The N-terminal pair of acidic residues (in wild-type yeast actin) determines the contribution of this site to weak and strong binding to S1. Additional acidic residues at this site (as shown by the 4Ac mutant and also implicated by α-actin; Miller et al. 1996b) do not improve binding, but probably assist in the isomerization of S1, increasing the ATPase, and hence the flux of cross-bridges through the cycle.

5 Specificity of Actomyosin Weak Binding

The results of studies with acidic pairs of residues on yeast actin show that as the ionic strength is increased in assay conditions, weak actomyosin binding decreases to a point at which the individual contribution of any of the three pairs of charged residues becomes essential for continued interaction, even in the presence of methyl cellulose. Conversely, when electrostatic interactions between actin and myosin are enhanced by decreasing the ionic strength and addition of methyl cellulose, the loss of any one of the acidic pairs does not effect protein interaction, and consequently, motility. This lack of functional resolution between the three pairs of acidic residues suggests that it is the net charge density of subdomain 1, rather than a specific single set of charged residues that determines whether the myosin progresses to the force producing states of the cross-bridge cycle. To test this conclusion more directly, a new pair of mutants was constructed in which the number of acidic residues on the surface of actin that are involved in myosin binding was unchanged, but their location within the three sites 24/25, 99/100, and N-terminus was altered (Wong et al. 1999). Thus, chimeric mutants were constructed in which either the D24A/D25A or E99A/E100A actin had additional acidic residues added to the N-terminus, as in the 4Ac mutant. The mutants 4Ac/D24A/D25A and 4Ac/E99A/E100A have the same number of acidic residues as each other, and also as the wild-type yeast actin, but each of the three actins lacks a pair of charged residues at one of the three locations. The overall charge density of these three proteins is thus the same, but the charge distribution on subdomain 1 is altered.

The results of all assays performed on these three proteins: binding to S1, actin-activated S1 ATPase, and in vitro motility assays, were indistinguishable from each other. Even scrutiny of the forces generated by the three actins, by in vitro motility assays with a load applied by modified HMM, revealed no differences between them (Wong et al. 1999). Thus, it is the overall charge density of subdomain 1 that dominates the weak interactions of actomyosin, rather than the specific location of these charges, and each pair of charged residues contributes equally to the weak actomyosin interaction.

6 Hydrophobic Strong Binding Residues

Residues involved in weak binding interactions have thus been extensively studied, and have in turn implicated one pair of these residues, at the N-terminus of actin, in enhancing the rigor binding of S1 to actin. Actin-myosin modeling predicts that other residues are involved in hydrophobic strong binding, specifically the α-helix 338–348 (Rayment et al. 1993b; Schroder et al 1993). This helix has a number of partially exposed hydrophobic residues including Ile341, Ile345 and Leu346, as well as the buried Trp340 residue. Two of these residues, I341 and I345, were chosen as targets for mutagenesis to alanines to investigate their role in rigor binding of S1 (Miller et al. 1996a). Substitutions of isoleucine to alanine were expected to leave the structure of actin unchanged, and yet affect the hydrophobic nature of these sites. Attempts to generate a double mutant of these sites resulted in a nonfunctional protein, evinced by non-viability of the yeast strain, as did attempts to increase the hydrophobicity of these residues by changing them to phenylalanines. However, these latter mutations may create steric problems for actin folding, hence leading to a non-functional protein.

The two mutant proteins, I341A and I345A, showed normal actin characteristics in polymerization and circular dichroism studies, confirming that these actins have no structural defects. The I345A mutant showed identical myosin binding properties, in vitro motility and actin-activated ATPase activ-

Table 4. Function parameters of hydrophobic actin mutants I341 A and I345A[a]

Functional parameter	Wild type/I345A	I341A
K_a (weak binding)	$1.8 \pm 0.3 \times 10^4\,M^{-1}$	$1.8 \pm 0.3 \times 10^4\,M^{-1}$
K_a (strong binding)	$1.5 \pm 0.2 \times 10^6\,M^{-1}$	$1.6 \pm 0.5 \times 10^5\,M^{-1}$
V_{max} (ATPase)	$2.4 \pm 0.3\,s^{-1}$	$0.6 \pm 0.1\,s^{-1}$
K_m (ATPase)	$15.0 \pm 6.8\,\mu M$	$10.5 \pm 5.4\,\mu M$
V_s (sliding speed in motility assays)	$2.5 \pm 0.4\,\mu m/s$	$1.6 \pm 0.4\,\mu m/s$

[a] Data taken from Miller et al. (1996a).

ities to the wild-type protein, suggesting no specific role in actomyosin function of this side chain. The I341A mutant, however, showed significant changes, specifically in the rigor complex, but not in the weakly bound complex with myosin. In the presence of ATP, binding of each of the actins to myosin was identical (Table 4). Rigor binding, in the absence of ATP, was reduced ninefold for the I341A actin compared to either wild-type or I345A actins (Table 4). Thus, I341 has an important role in strong actomyosin binding, but not weak binding.

Mutagenesis of Ile341 also affected actin-activated ATPase of S1, specifically decreasing the V_{max} fourfold compared with other actins (Table 4). The K_m, however, showed no significant difference among the three proteins. Together, these results implicate Ile341 in the catalytic interaction between actin and myosin, but not in the binding of S1 in the presence of ATP. Thus, in addition to its effect on the strong binding of actin to myosin, the I341A mutation most likely affects the isomerization between the $AM^{**}\cdot ADP\cdot P_i$, and $AM'\cdot ADP\cdot P_i$ states in the cross-bridge cycle.

The I341A mutation was also found to decrease the sliding speed of the actin filaments, with the I345A mutant again being identical to wild-type yeast actin (Table 4). Decreased force generation with the I341A actin was confirmed using NEM-modified HMM as an external load in the in vitro motility assays. The relative force generated by the I341A actin was reduced by 40% compared with the wild-type actin. In principle, the reduced force could result from a smaller force generated by a given cross-bridge, or a decreased number of force-generating cross-bridges in the mutant actomyosin ATPase cycle. The latter possibility is consistent with the decreased transition of I341A actin from weakly to strongly bound complexes and the overall decrease in the strong binding of S1 to this actin. The fact that this mutation has virtually no effect on the weak actomyosin interaction is apparent also from the insensitivity of the motility of this actin to such factors as the presence of methyl cellulose or high salt concentrations in the in vitro motility assays. Clearly, the I341A mutant is distinct from the weak binding charge mutants in its myosin binding properties, and consequently its mechanical behavior.

7 C-Terminus of Actin and the 262–274 Plug

The myosin light chain (A1) binding site on actin, involving residues 262–274 implicated in such binding via carbodiimide cross-linking experiments (Sutoh 1983), has also been examined with the help of *Dictyostelium* E360, E361, E363 and E364 actin mutants (Sutoh et al. 1991). In general, the results of this study are consistent with the proposed interaction of the above residues with the A1 light chain, but the critical tests of such binding have yet to be performed. If, as proposed, 360–364 segment is the dominant binding site for A1, then the main biochemical difference between S1 isozymes (Wagner at al. 1979) and in vitro motility differences between myosin A1 and myosin A2 complexes (Lowey et al. 1993) with actin should be reduced substantially or canceled when using the above mutants. Clearly, it would be important to carry out such experiments.

Two mutations in the C-terminal residues of muscle actins have been identified in different organisms that result in altered muscle physiology. A mutation in the *Drosophila melanogaster* flight muscle *Act88F* gene G368E causes flight impairment, with a wing-beat frequency reduced by about 30% (Anson et al. 1995). A more recent report has identified a pair of actin mutations that are responsible for idiopathic dilated cardiomyopathy (IDC) in humans (Olson et al. 1998). One of these mutations (R312H) is at a residue that has been suggested to play a role in tropomyosin/troponin regulation of the muscle (Lorenz et al. 1995); a potential mechanism of action of the other mutation (E361G) has not been identified. Although the C-terminus of actin is in subdomain 1, the crystallographic structure reveals that both of these mutations (G368E in *Act88F* and E361G in cardiac actin) are at the back of the actin molecule, potentially buried in the filament structure. One of these residues, E361, has been implicated in the binding of the myosin light chain (Sutoh 1982). In studies with the partially purified fly protein (the purification yielded a mixture of different actins, and the data implicates differential myosin interactions with the mutant actin), a slight decrease in the in vitro motility speeds was seen, and significant differences in myosin/actin binding were observed, with a threefold increase in K_d. Actin-activated ATPase was not altered by the mutation (Anson et al. 1995). It was thus concluded that the G368E mutation affects the ability of myosin to bind actin in the weakly bound pre-force generating step, or its ability to isomerize into the strongly bound state. Biochemical analysis has not been performed on the cardiomyopathy mutant actin (E361G), and it will be interesting to see if similar changes occur with this protein.

It is interesting to note that four other proteins, cofilin (McGough et al. 1997), fimbrin (Hanein et al. 1997), gelsolin (McLaughlin et al. 1993) and scruin (Owen and DeRosier 1993) for which binding to actin filaments has been modeled, all induce conformational changes in the filament structure at the subdomain of actin where each primarily binds. Although cofilin and gelsolin destabilize the actin filament and lead to its cleavage, fimbrin and scruin stabilize actin filaments and induce their bundling by binding two adjacent filaments. It has been suggested that fimbrin induces a small rotation of subdomain 1 of actin upon binding (Hanein et al. 1997). It does not seem unlikely, therefore, that myosin might also induce a similar change in the actin filament upon binding (Egelman and Orlova 1995), and that the G368 and E361 residues play a role in this movement. It has already been noted that myosin binding and actin activated ATPase increase in a co-operative fashion with actin previously bound to an antibody to the N-terminal seven residues of actin [$F_{ab(1-7)}$; DasGupta and Reisler 1992]. Might this behavior be due to a conformational change in the actin filament upon binding myosin?

Intriguing evidence on myosin-induced changes, spreading as far as the intrastrand contacts in actin filaments, has been obtained from the cysteine mutant S265C (with C374A) in the hydrophobic plug 262–274 of yeast actin (Feng et al. 1997). The main role of this plug has been postulated to be the stabilization of F-actin structure by providing the lock and key type contacts between monomers on opposite strands of filaments (Lorenz et al. 1993). Despite some distance from the myosin binding surface on actin, the fluorescence of pyrene-labeled S265 C actin is increased by ~25% by S1 (Feng et al. 1997). This observation, in addition to documenting myosin-caused changes in the filament interior, focuses attention on the power and promise of combined mutational and spectroscopic approaches in mapping those regions on actin which change upon binding of other proteins. This appears to be an interesting and important avenue for future investigations.

8 Conclusions

Mutational studies on actin have clearly increased our understanding of actomyosin interactions. Primarily, mutant actins, if viable for cell growth, provide us with proteins with minimal structural perturbations. Amino acid changes can be engineered to investigate precise interactions, whilst maintaining the overall structure and function, and subtly changing the characteristics of the residue of interest (e.g. charge, size, or hydrophobicity). These mutants offer obvious advantages over chemical modifications of actin which usually result in the addition of bulky groups, that may alter interactions with myosin, in direct and indirect ways (Crosbie et al. 1994). Similarly, binding of antibodies to actin also introduces bulky groups to the actin filament which cause steric interference to myosin interaction. Both of these approaches also depend on the reactivity (and specificity) of the site of attachment and, therefore, can be seldom targeted to the desirable regions on the protein.

The current model of the actomyosin complex based on docking the S1 structure to F-actin (Rayment et al. 1993b; Schröder et al. 1993) is thought to represent these proteins in the rigor state of the cross-bridge cycle. Although interaction residues on both actin and myosin have been identified in this model, as well as in earlier biochemical and immunological studies, a greater understanding of the weak and strong binding to myosin has emerged from

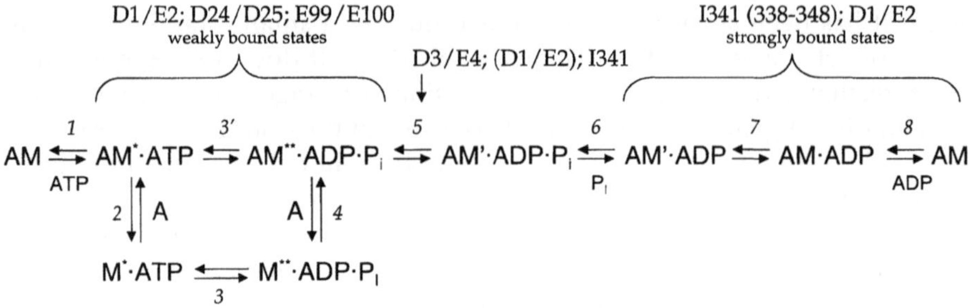

Fig. 2. Schematic representation of actomyosin cross-bridge cycle. *A* Actin; *M*, *M**, *M***, and *M′* different conformations of myosin

the use of actin mutants. The cross-bridge cycle involves a number of steps (see Fig. 2), and the importance of different residues on actin in these steps has been elucidated by the mutational work. The charged residues D1/E2, D24/D25, and E99/E100 have been identified as playing a role in the weak binding of myosin to actin (Sutoh et al. 1991; Cook et al. 1992; Johara et al. 1993; Miller and Reisler 1995; Miller et al. 1996b). The results from studies with the N-terminal mutants, however, revealed their role in an additional step, the transition from weakly to strongly bound states (step 5 in Fig. 2). This is the rate limiting step of the actin-activated ATPase, and involves the transition from the low affinity actin·myosin·ADP·P_i (A·M**·ADP·P_i) to an activated, higher affinity A·M′·ADP·P_i complex (conformationally strained, ready for the power stroke). The N-terminus of actin facilitates this transition by interacting with the lysine-rich 626–647 rich loop on myosin. This is evident from the results with the 4Ac yeast actin mutant (Cook et al. 1993; Miller et al. 1996b), and is supported by a complementary study of Furch et al. (1998) in which the 626–647 loop on myosin has been replaced with constructs of different size and charge density. The weak and strong myosin binding affinity of both 4Ac and wild-type actin are identical, and thus the increased V_{max} of the 4Ac acto-S1 ATPase must stem from an improved isomerization between A·M**·ADP·P_i, and A·M′·ADP·P_i (step 5) for the 4Ac protein. This, in turn, translates into improved flux of cross-bridges through the cycle, an increased population of strongly bound states and, therefore, greater force generated with HMM than that produced by wild-type actin.

Charge mutants DNEQ, ΔDSE, D24A/D25A, and E99A/E100A all show decreased force generation in the in vitro motility assays, whilst having rigor binding to S1 identical with wild-type actin, so the decrease in the population of strongly bound (and hence force-generating) cross-bridges for these actins must arise from a smaller population of weakly bound complexes. Thus, fewer myosin cross-bridges reach the isomerization step (step 5) with mutant actins. In addition, these mutants also display a decreased V_{max} for actin-activated S1-ATPase, indicating some decrease in the isomerization and thus, a further decrease in the flux of cross-bridges through the cycle into strongly bound states. Similar analysis of the hydrophobic mutant I341A (Miller et al. 1996a), with its decreased motility and force generation, shows that its functional impairment occurs at a step subsequent to the weak binding equilibria (steps 1–4 in Fig. 2). On the basis of the changes in V_{max} and rigor S1 binding, I341A affects the isomerization (step 5) and strong binding equilibria steps in the cycle. A reduced force is most likely produced because of changes in these steps in the cross-bridge cycle.

Clearly, the use of actin mutants has helped dissect acto-S1 contacts in the different stages of the cross bridge cycle, and identify residues important for the transition between some steps. Evidence for myosin-induced changes in actin has been obtained with such mutants (Feng et al. 1997), and new insight into regulation of actomyosin interactions begins to emerge from mutational studies on actin (Saeki et al. 1996; Bing et al. 1998). However, it is clear that the mapping of the myosin interface on actin is still incomplete, and residue-detailed information on actin dynamics and actin regulation is yet to be developed. In addition, the role of specific residues on the myosin molecule in the cross-bridge cycle has only started to be investigated in this fashion. Mutations of both proteins, and their analysis against their wild-type and mutated partners, culminating perhaps in the generation of myosin suppressors of actin mutants, and vice versa, will further advance this fascinating field.

Acknowledgements. Awards from the USPHS (AR22031) and the NSF (MCB9630977) supported this work.

References

Adams S, Reisler E (1993) Role of sequence 18–29 on actin in actomyosin interactions. Biochemistry 32:5051–5056

Anson M, Drummond DR, Geeves MA, Henessey ES, Ritchie MD and Sparrow JC (1995) Actomyosin kinetics and in vitro motility of the wild-type *Drosophila* actin and the effects of two mutations in the *Act88F* gene. Biophys J 68:1991–2003

Aspenström P, Karlsson R (1991) Interference with myosin subfragment-1 binding by site-directed mutagenesis of actin. Eur J Biochem 200:35–41

Bass SH, Mulkerrin MG, Wells JA (1991) A systematic mutational analysis of hormone-binding determinants in the human growth hormone receptor. Proc Natl Acad Sci USA 88:4498–4502

Bennett WF, Paoni NF, Keyt BA, Botstein D, Jones AJ, Presta L, Wurm FM, Zoller MJ (1991) High resolution analysis of functional determinants on human tissue-type plasminogen activator. J Biol Chem 266:5191–5201

Bing W, Razzaaq A, Sparrow J, Marston S (1998) Tropomyosin and troponin regulation of wild type and E93 K mutant actin filaments from *Drosophila* flight muscle. J Biol Chem 273:15016–15021

Cook RK, Sheff DR, Rubenstein PA (1991) Unusual metabolism of the yeast actin amino terminus. J Biol Chem 266:16825–16833

Cook RK, Blake WT, Rubenstein PA (1992) Removal of the amino-terminal acidic residues of yeast actin. J Biol Chem 267:9430–9436

Cook RK, Root D, Miller C, Reisler E, Rubenstein PA (1993) Enhanced stimulation of myosin subfragment 1 ATPase activity by addition of negatively charged residues to the yeast actin NH2 terminus. J Biol Chem 268:2410–2415

Cooke R (1997) Actomyosin interaction in striated muscle. Physiol Rev 77:671–697

Crosbie RH, Miller C, Cheung P, Goodnight T, Muhlrad A, Reisler E (1994) Structural connectivity in actin – effect of C-terminal modifications on the properties of actin. Biophys J 67:1957–1964

Cunningham BC, Wells JA (1989) High resolution epitope mapping of hGH-receptor interactions by alanine-scanning mutagenesis. Science 244:1081–1085

DasGupta G, Reisler E (1989) Antibody against the amino terminus of α-actin inhibits actomyosin interactions in the presence of ATP. J Mol Biol 207:833–836

DasGupta G, Reisler E (1991) Nucleotide-induced changes in the interaction of myosin fragment 1 with actin: detection by antibodies against the N-terminal segment of actin. Biochemistry 30:9961–9966

DasGupta G, Reisler E (1992) Actomyosin interactions in the presence of ATP and the N-terminal segment of actin. Biochemistry 31:1836–1841

dos Remedios CG, Moens PDJ (1995) Actin and the actomyosin interface: a review. Biochim Biophys Acta 1228:99–124

Egelman EH, Orlova A (1995) New insights in actin filament dynamics. Curr Opin Struct Biol 5:172–180

Feng L, Kim E, Lee W-L, Miller CJ, Kuang B, Reisler E, Rubenstein PA (1997) Fluorescence probing of yeast actin subdomain 3/4 hydrophobic loop 262–274. J Biol Chem 272:16829–16837

Furch M, Geeves MA, Manstein DJ (1998) Modulation of actin affinity and actomyosin adenosine triphosphatase by charge changes in the myosin motor domain. Biochemistry 37: 6317–6326

Gallwitz D, Suers I (1980) Structure of a split yeast gene: Complete nucleotide sequence of the actin gene in *Saccharomyces cerevisiae*. Proc Natl Acad Sci USA 77:2546–2550

Hanein D, Matsudaira P, DeRosier DJ (1997) Evidence for a conformational change in actin induced by fimbrin (N375) binding. J Cell Biol 139:387–396

Holmes KC (1995) The actomyosin interaction and its control by tropomyosin. Biophys J 68: S2-S7

Holmes KC, Popp D, Gebhard W, Kabsh W (1990) Atomic model of the actin filament. Nature 347:44–49

Johannes F-J, Gallwitz D (1991) Site-directed mutagenesis of the yeast actin gene: a test for actin function in vivo. EMBO J 10:3951–3958

Johara M, Toyoshima YY, Ishijima A, Kojima H, Yanagida T, Sutoh K (1993) Charge-reversion mutagenesis of *Dictyostelium* actin to map the surface recognized by myosin during ATP-driven sliding motion. Proc Natl Acad Sci USA 90:2127–2131

Kabsh W, Mannherz HG, Suck D, Pai EF, Holmes KC (1990) Atomic structure of the actin:DNase I complex. Nature 347:37–44

Karlsson R, Aspenström P, Byström AS (1991) A chicken β-actin gene can complement a disruption of the *Saccharomyces cerevisiae ACT1* gene. Mol Cell Biol 11:213–217

Kim E, Miller CJ, Motoki M, Seguro K, Mulrad A, Reisler E (1996) Myosin-induced changes in F-actin: fluorescence probing of subdomain 2 by dansyl ethylenediamine attached to Gln-41. Biophys J 70:1439–1446

Kron SJ, Drubin DG, Botstein D, Spudich J (1992) Yeast actin filaments display ATP-dependent sliding over the surfaces coated with rabbit muscle myosin. Proc Natl Acad Sci USA 89: 4466–4470

Lorenz M, Popp D, Holmes KC (1993) Refinement of the F-actin model against X-ray fiber diffraction data by use of a directed mutation algorithm. J Mol Biol 234:826–836

Lorenz M, Poole KJV, Popp D, Rosenbaum G, Holmes KC (1995) An atomic model of the unregulated thin filament obtained by X-ray fiber diffraction on oriented actin-tropomyosin gels. J Mol Biol 246:108–119

Lowey S, Waller GS, Trybus KM (1993) Function of skeletal muscle myosin heavy and light chain isoforms by an in vitro motility assay. J Biol Chem 268:20414–20418

McGough A, Pope B, Chiu W, Weeds A (1997) Cofilin changes the twist of F-actin: implications for actin filament dynamics and cellular function. J Cell Biol 138:771–781

McLaughlin PJ, Gooch JT, Mannherz H-G, Weeds AG (1993) Structure of gelsolin segment 1-actin complex and the mechanism of filament severing. Nature 364:685–692

Miller CJ, Reisler E (1995) Role of charged amino acid pairs in subdomain-1 of actin in interactions with myosin. Biochemistry 34:2694–2700

Miller L, Kalnoski M, Yunossi Z, Bulinski JC, Reisler E (1987) Antibodies directed against N-terminal residues on actin do not block acto-myosin binding. Biochemistry 26:6064–6070

Miller CJ, Cheung P, White P, Reisler E (1995) Actin's view of actomyosin interface. Biophys J 68:S50–S54

Miller CJ, Doyle TC, Bobkova E, Botstein D, Reisler E (1996a) Mutational analysis of the role of hydrophobic residues in the 338–348 helix on actin in actomyosin interactions. Biochemistry 35:3670–3676

Miller CJ, Wong WW, Bobkova E, Rubenstein PA, Reisler E (1996b) Mutational analysis of the role of the N terminus of actin in actomyosin interactions. Comparison with other mutant actins and implications for the cross-bridge cycle. Biochemistry 35:16557–16565

Ng R and Abelson J (1980) Isolation of the gene for actin in *Saccharomyces cerevisiae.* Proc Natl Acad Sci USA 77:3912–3916

Olson TM, Michels VV, Thibodeau SN, Tai YS, Keating MT (1998) Actin mutations in dialated cardiomyopathy, a heritable form of heart disease. Science 280:750–752

Owen C, DeRosier D (1993) A 13-Å map of the actin-scruin filament from the limulus acrosomal process. J Cell Biol 123:337–344

Rayment I, Rypniewski WR, Schmidt-Bäse K, Smith R, Tomchick DR, Benning MM, Winkelmann DA, Wesenberg G, Holden HM (1993a) Three-dimensional structure of myosin subfragment-1: a molecular motor. Science 261:50–58

Rayment I, Holden HM, Whittaker M, Yohn CB, Lorenz M, Holmes KC, Milligan RA (1993b) Structure of actin-myosin complex and its implications for muscle contraction. Science 261:58–65

Saeki K, Sutoh K, Wakabayashi T (1996) Tropomyosin-binding site(s) on the *Dictyostelium* actin surface as identified by site-directed mutagenesis. Biochemistry 35:14465–14472

Schröder RR, Manstein DJ, Jahn W, Holden H, Rayment I, Holmes KC, Spudich JA (1993) Three-dimensional atomic model of F-actin decorated with *Dictyostelium* myosin. Nature 364:171–174

Sheterline P, Sparrow JC (1994) Actin. Protein Profile 1:1–121

Sutoh K (1982) Identification of myosin-binding sites on actin sequence. Biochemistry 21: 3654–3661

Sutoh K (1983) Mapping of actin-binding sites on the heavy chain of myosin subfragment 1. Biochemistry 22:1579–1585

Sutoh K, Ando M, Sutoh K, Toyoshima YY (1991) Site-directed mutations of *Dictyostelium* actin: disruption of a negative charge cluster at the N-terminus. Proc Natl Acad Sci USA 88:7711–7714

Wagner PD, Slater CS, Pope B, Weeds AG (1979) Studies on the actin activation of myosin subfragment-1 isozymes and the role of myosin light chains. Eur J Biochem 99:385–394

Wertman KF, Drubin DG, Botstein D (1992) Systematic mutational analysis of the yeast *ACT1* gene. Genetics 132:337–350

Wong WW, Doyle TC, Reisler E (1999) Nonspecific weak actomyosin interactions: relocation of charged residues in subdomain 1 of actin does not alter actomyosin function. Biochemistry 38:1365–1370

Role of Charges in Actomyosin Interactions

Patrick Chaussepied and Juliette Van Dijk[1]

1 Introduction

The interaction of myosin with actin is of great interest to researchers working on cell motility processes because it participates actively in the generation of actin based movement. But this interaction presents a more general interest for protein workers since its formation is a multistep binding process strongly regulated by effectors such as ATP derivatives and regulatory systems. As with most protein–protein interactions, the actomyosin complex presents a composite interface made of multiple electrostatic and hydrophobic contacts. The goal of this chapter is to identify the electrostatic contacts, to describe their role during the formation of the actomyosin interface and to discuss their contribution in the mechanism of force generation.

2 Structure of the Actomyosin Interface

The atomic structures of the actin monomer (Kabsch et al. 1990), of the actin filament (F-actin; Lorenz et al. 1993) and of the myosin motor domain (also named subfragment 1 or S1; Rayment et al. 1993a) are available. However, there is no structural information on the active filamentous actomyosin complex at very high resolution, i.e. based on the crystal structure of actin bound to either myosin or its active motor domain. This is due primarily to the polymerized state of the actin filament that prevents its crystallization. In addition, the unpolymerized complex, formed by mixing monomeric actin with myosin S1, is not stable and does not have normal ATPase activity (Lheureux et al. 1993; Lheureux and Chaussepied 1995a). An explicit model of the actomyosin interface has been obtained using the X-ray structures of the individual proteins together with data of fiber diffraction and electron microscopy of actin filaments saturated by S1 (Rayment et al. 1993b; Schröder et al. 1993; Mendelson and Morris 1997). Three recent reviews sum up our knowledge on the structure of the actin-myosin interface and discuss the validity of the different experimental approaches used (Dos Remedios and Moens 1995; Milligan 1996; Cooke 1997).

[1] CRBM du CNRS, IFR24, 1919 route de Mende, 34293 Montpellier Cedex 5, France

Results and Problems in Cell Differentiation, Vol. 36
D.D. Thomas, C.G. dos Remedios (Eds.): Molecular Interactions of Actin

The interface is defined by four main subsites with various electrostatic and hydrophobic components (Fig. 1). The primary and the most studied electrostatic subsite encompasses a patch of acidic residues located on actin subdomain 1 (subsite 1 composed of actin residues 1–4, 24–25 and 99–100) and a lysine-rich loop of S1 (residues 626–647 also called loop 2, numbers are for skeletal muscle myosin). The interface is also composed of multiple hydrophobic and stereospecific contacts located in actin, mainly between subdomains 1 and 3 (subsites 2a and 3) and within loop 40–52 of subdomain 2 (subsite 2b) of the adjacent monomer and in S1, between residues 529–558 (subsite 2) and 403–416 (subsite 3). Only a few experimental studies have characterized these apolar interactions (Onishi et al. 1995; Patterson and Spudich 1996; Bertrand et al. 1997).

Among the other probable components of the interface, electron microscopy reconstruction suggested a secondary electrostatic site between actin residues 99–100 and S1 loop 567–578 (named loop 3). This secondary site is apparently weak in the saturated 1/1 complex since it was not confirmed by recent re-analysis of the electron microscopy images (Mendelson and Morris 1997) nor by chemical cross-linking experiments (Andreev and Borejdo 1995). However, this secondary site reinforces under non-saturating conditions in which both loop 2 and loop 3 from skeletal or cardiac muscle myosin can be cross-linked with the segment 1–12 of two adjacent actin monomers (Andreev and Borejdo 1992; Bonafé and Chaussepied 1995; Andreev and Borejdo 1997). In order to fit this site in the 3-D map of the complex, one had to re-orient (though to a small extent) S1 over the filament axes. Such a change in orientation of the myosin head was in fact observed by fluorescence polarization studies when saturated and unsaturated filaments were compared in muscle fibers or in myofibrils (Andreev et al. 1993; Xiao and Borejdo 1997). These authors pro-

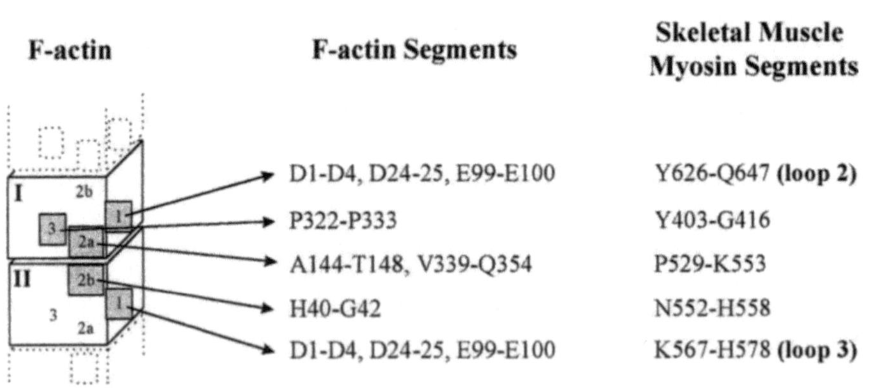

Fig. 1. Schematic representation of the skeletal muscle myosin binding sites on F-actin. Two actin monomers (*I* and *II*) of the same strand of the filament are drawn. The contact sites of one myosin motor domain: *1*, *2a*, *3* and *2b* are *highlighted*. See text for more details

posed that this re-orientation could take place at the end of the power stroke in the process of force generation.

It appears that the two main electrostatic contacts involve two positive loops protruding from the surface of the myosin motor domain (Diaz Banos 1996) and the same patch of negative residues on two different actin monomers. It is interesting to note that, among the different actin and myosin isoforms, the primary sequences of these loci are highly variable, both in charge content and in length (Sheterline and Sparrow 1994; Cope et al. 1996). It was proposed that this sequence variability could be linked to the difference in catalytic efficiency of the various isotypes of actomyosin complexes (Uyeda et al. 1994; Furch et al. 1998).

3 Dynamics of the Actomyosin Complex

Myosin movement along F-actin is due to cyclic changes in the interaction of S1 with actin, under the control of ATP hydrolysis (Lymn and Taylor 1971). Kinetic studies on the actin-myosin ATPase reaction (Bagshaw and Trentham 1974; Eisenberg and Hill 1985) and on the binding of myosin to actin (Geeves and Conibear 1995) have suggested a common kinetic model for all myosin isoforms with slight differences in kinetic constants (Ostap and Pollard 1996). This model starts at the beginning of the power stroke when actin binds weakly to S1 containing ATP or the hydrolyzed ADP.Pi intermediate in its active site. The resulting transient A-S1.ATP or A-S1.ADP.Pi complexes are related to the collision complexes characterized by long-range electrostatic interactions (Highsmith and Murphy 1992). The following isomerization of S1.ADP.Pi complex and/or the release of the γPi results in the force generating A-S1′.ADP.Pi or A-S1′.ADP complex (Hibberd and Trentham 1986; Homsher et al. 1997). Although this last A-S1′-ADP(.Pi) intermediate has not yet been isolated, it should be analogous to an attached low-affinity state (A-state), composed of both hydrophobic and ionic bonds (Geeves and Conibear 1995). Finally, a second isomerization, which corresponds to the release of Pi and ADP, generates the A-S1.ADP and actin-S1 complexes. These complexes, characterized by high affinities and low salt dependencies, predominate in the so-called strong or rigor conformational state (R-state; Chaussepied et al. 1988; Highsmith 1990). Force would be developed during the A-state to R-state transition of the actin-S1 interface. A major goal of current research in the field is to identify the structure of the actin-myosin interface for each kinetically defined state (Cooke 1997).

4 Role of the Ionic Interactions

The first contacts between two proteins, which result in the so-called collision complex, take place almost exclusively via coulombic attractions. Indeed, these contacts involve non-specific charge–charge, charge–dipole and dipole–dipole interactions (Schreiber and Fersht 1996). Due to the short half-life of the collision complex, it is usually not possible to identify the components of its interface unless it is artificially stabilized, for example by inhibiting the subsequent contacts needed for stereospecific interactions (see below).

F-actin possesses a rather strong electric dipole moment (Kobayashi et al. 1964; Cantiello et al. 1991) that may be involved in the capture of the positively charged loops present at the surface of the myosin motor domain (Diaz Banos et al. 1996). Interestingly, the same myosin loops (loop 2 and loop 3) would be involved in both the non-specific collision contacts and in the stereospecific final interface. However, as we will describe below, the relative importance of the two loops differs in these two types of complexes.

Numerous works using biochemical, immunological and directed mutagenesis approaches (reviewed in Van Dijk et al. 1998) clearly demonstrated that actin binding to loop 2 is essential for the actomyosin ATPase activity and more specifically for the formation of the collision (also called weak binding) complex. Recent methodical mutagenesis experiments coupled with rapid kinetic studies have revealed that the number of charges (rather than the length) of loop 2 is essential for its optimal function (Furch et al. 1998). It was also clearly established that loop 2 is of minor importance in stabilizing the rigor interface (Chaussepied 1989; Das Gupta and Reisler 1992; Bobkov et al. 1996; Miller et al. 1996).

The role of the second electrostatic binding site located in myosin loop 3 is more controversial. There are four arguments against this site having an active role in the actomyosin complex. Firstly, the number of positively charged residues in loop 3 is highly variable among the members of the myosin family (Sellers and Goodson 1995). Secondly, loop 3 is thought to interact with actin almost exclusively when actin is in excess of S1, i.e. when S1 can bind freely to two adjacent actin monomers. Thirdly, it was shown that the degree of saturation of the filament by S1, which modulates actin binding to loop 3, does not affect the ATPase activity (Lheureux and Chaussepied 1995b). Fourthly, antibodies directed against loop 3 do not significantly inhibit the formation of the actin-myosin complex (Blotnick et al. 1995). However, the same antibodies strongly reduce the sliding velocity of actin filaments in in vitro motility assays, in total agreement with an essential function of this loop during the force generation process. The data presented below suggests that the myosin loop 3 may indeed participate directly in the formation of the weak binding interface, at least in the case of the skeletal muscle actin-myosin complex.

5 Studies of the Ionic Interactions by Chemical Cross-Linking Experiments

The electrostatic contacts within protein–protein complexes are obviously best studied by methods that provide structural information such as NMR and XR crystallography. But these methods are limited by the size of the complexes or by the high flexibility of segments which, like myosin loop 2 and loop 3, cannot be seen in the 3-D structure. Among the experimental approaches that can overcome these limitations, one finds specific chemical or genetic modification, immunochemistry or peptide-based competition experiments and chemical cross-linking studies. This last approach presents, however, several advantages. It is the most efficient way to trap (or to freeze) transient interfaces without disturbing their formation. Moreover, it could, using flash-controlled photo-cross-linking experiments, identify nanosecond-scale changes in protein–protein interactions. As we shall describe below, cross-linking experiments have given a preliminary explicit idea on the modification of the ionic contacts during the mechanochemical cycle (Van Dijk et al. 1998).

5.1 Cross-Linking Reactions and Identification of the Cross-Linking Sites

A large number of cross-linking reagents have been used on the actin-myosin complex (Mornet et al. 1981; Labbé et al. 1982; Bertrand et al. 1988, 1997; Bonafé et al. 1993). The one that provided the most interesting information on the coulombic interface was the 1-ethyl-3[3-(dimethyl amino)-propyl]-carbodiimide (EDC) that forms isopeptidic bonds where ionic interactions take place. EDC has not only been extensively used to characterize the actin-myosin interface with proteins in solution (Mornet et al. 1981; Chen et al. 1985; Craig et al. 1985; Arata 1986; Yamamoto 1989; Andreev and Borejdo 1992; Bonafé and Chaussepied 1995) but also within myofibrils or muscle fibers (Tawada and Kimura 1986; Tawada and Kawai 1990; Duong and Reisler 1989; Herrmann et al. 1993). EDC-induced reaction on the actin-S1 complex generated three main cross-linking products of 165, 175 and 265 kDa and a minor product of 200 kDa. Identification of the cross-linking sites on actin and myosin was accomplished by both chemical and proteolytic degradation of the cross-linked complexes (Sutoh 1982, 1983; Bertrand et al. 1988; Combeau et al. 1992; Bonafé and Chaussepied 1995). The 165-, 175- and 200-kDa products contain only one actin bound to one S1, whereas the 265-kDa product was formed by two actin monomers attached to the same S1 molecule. In all cases, cross-linking occurred within the N-terminal actin segment 1–12 (subsite 1 in Fig. 2). The different masses obtained for the 165-kDa and 175-kDa products were explained by actin cross-linking to two different myosin loci, loop 2 or loop 3. The 265-kDa product resulted in the simultaneous linkage of loops 2 and 3 to the N-terminal segment of two adjacent actin monomers (Fig. 2). Only striated muscle myosin (from skeletal or cardiac muscle) and not smooth muscle myosin was found to cross-link via the loop 3 segment, thus generating the 175-kDa band and the doubly cross-linked 265-kDa product (Marianne-Pepin et al. 1985). The 200-kDa product, usually much less abundant, has not yet been clearly identified.

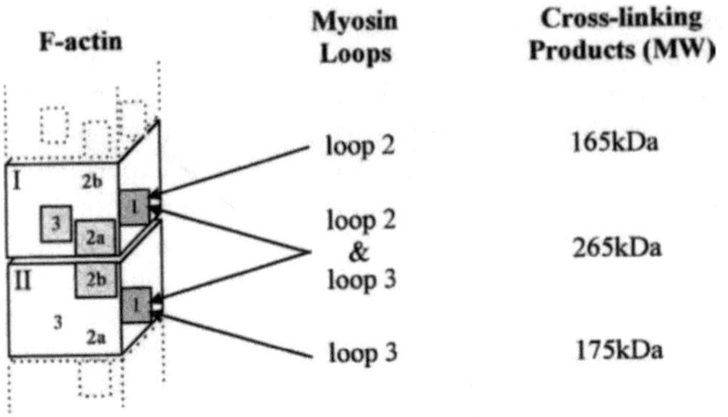

Fig. 2. Schematic drawing of the EDC-induced cross-linking sites. Cross-linking products of 165, 175 and 265 kDa correspond to the linkage between residues 1–12 of actin (site 1) with S1 loop 2 (residues 626–647), S1 loop 3 (residues 567–578), and both loop 2 and loop 3, respectively

One possible artifact with this type of reagent could occur from cross-linking of non-specific ionic interactions, such as those taking place during the formation of most of the protein–protein complexes (Janin 1997). However, this possibility can be rejected here since image reconstructions of the complex clearly show either direct involvement or close proximity in the interface of the two cross-linked counterparts (see also discussion in Van Dijk et al. 1998).

5.2 Regulation of the Cross-Linking Sites by Nucleotide Analogues

The lifetimes of the transient states in the actin myosin ATPase cycle are too short to allow chemical cross-linking reactions. Different nucleotide analogues are usually employed to freeze each critical conformational state. The "rigor-like" complexes are simply obtained in the absence of nucleotide or in the presence of ADP. ATP analogues, such as AMPPNP, ATPγS and ADP.BeFx induce similar "ATP-like" changes in the crystal structure of the catalytic domain of S1 (Gulick et al. 1997) though slight differences regarding other structural properties of the ternary actin-S1.nucleotide complex may be observed (Duong and Reisler 1989; Roopnarine and Thomas 1996). Finally, again based on crystallographic studies, ADP.AlF_4^- and ADP.VO_4^{3-} are employed to mimic a different S1.ADP.Pi state (Fisher et al. 1995; Smith and Rayment 1996).

The results of the cross-linking experiments performed on unsaturated complexes in the presence of various nucleotide analogues are reported in Fig. 3. Firstly, the pattern of the cross-linking time course is very similar in the absence and in the presence of ADP, with the appearance of 165- and 175-kDa products (in equal amounts) followed by an increase in the 265-kDa band. Secondly, with AMPPNP, ADP.BeFx, ADP.AlF_4^- and ADP.VO_4^{3-}, the 265-kDa band was virtually the only product detected on the gel, revealing that one molecule

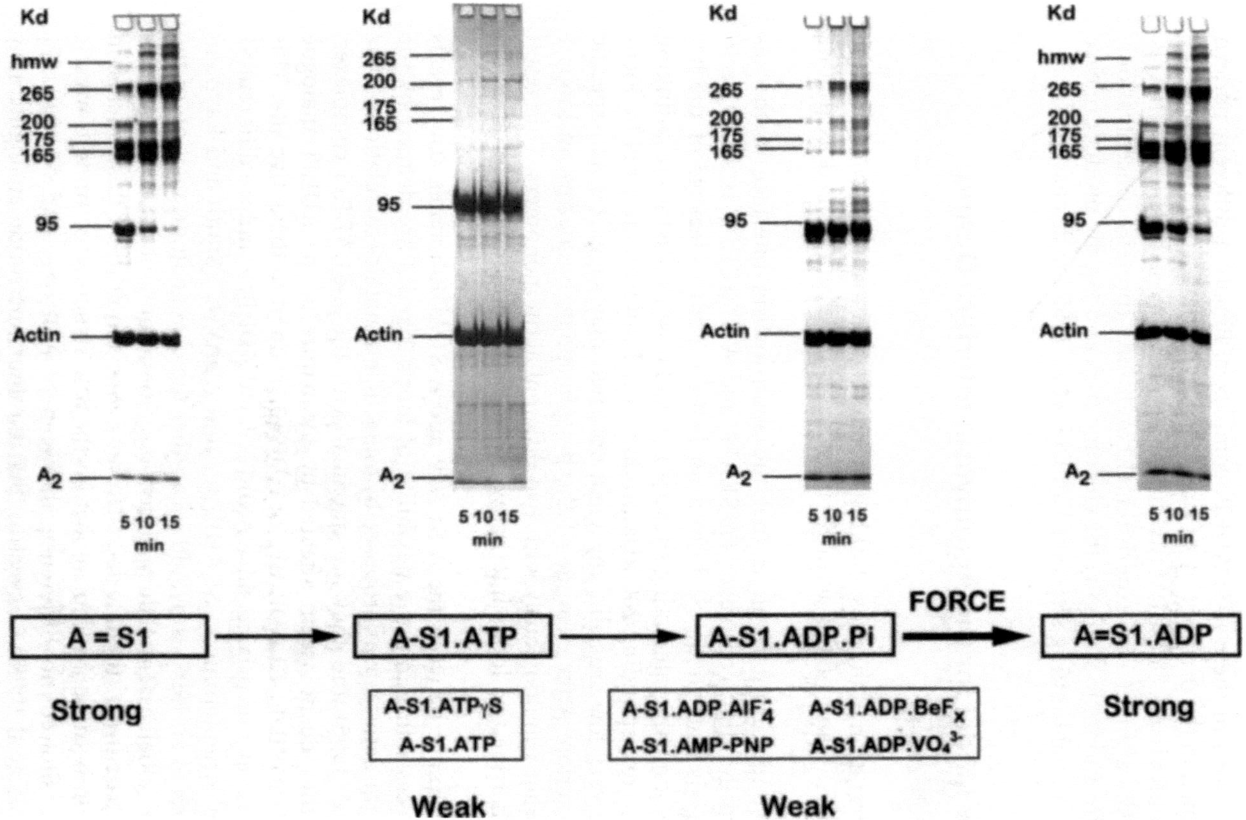

Fig. 3. Cross-linking experiments on actin-S1 complexes in various intermediate states. Cross-linking reactions were performed on a mixture of 45 μM actin and 15 μM S1 as described by Van Dijk et al. (1998). Quickly, the actin-S1.nucleotide complexes were mixed in pH 7.0 buffer (30 mM MOPS, 2.5 mM $MgCl_2$) with 15 mM EDC at 20 °C. At the times indicated, the reaction was stopped with the Laemmli buffer (Laemmli 1970) and analyzed on an SDS-polyacrylamide 4–18% gradient gel. S1 is composed of the 95-kDa heavy chain segment and of alkali light chain 2 (A2). The *numbers* are molecular masses in kDa

of S1 is cross-linked – i.e., interacts – simultaneously with the N-termini of two actin monomers in these A-S1.ADP.Pi complexes. Finally, a different pattern was generated in the presence of ATP or ATPγS, with the formation of only the minor 200-kDa band. These last two analogues are supposed to induce an ATP-like state since under our experimental conditions (protein concentrations, temperature and ionic strength) the cleavage step, and not the product release step, is the actual limiting step (see Van Dijk et al. 1998 for discussion).

6 A New Model for the Actomyosin Interface During the ATPase Cycle

The interpretation of the cross-linking data in relation to our knowledge of the structure of S1 and of the actin-S1 interface can be represented as follows (Fig. 4).

The first important result is that the actin-myosin interface in the so-called weak binding complexes (A-S1.ATP and A-S1.ADP.Pi) is sensitive to the hydrolysis state of the S1 bound nucleotide. Before cleavage of the β-γ phosphate bond, the interacting sites are not yet known since the cross-linking sites in the corresponding 200-kDa product could unfortunately not be identified. The very low cross-linking yield obtained may suggest a non-specific ionic interaction. After cleavage of the β-γ phosphate bond, myosin loops 2 and 3 interact simultaneously with two different actin monomers. This result suggests that the two loops interact with two actin monomers with an equal affinity.

The difference between the A-S1.ATP and A-S1.ADP.Pi states also reveals that the accessibility and the flexibility of loops 2 and 3 are defined by the content of the active site as proposed by molecular dynamics calculations (Diaz Banos et al. 1996) and they are maximal with the A-S1.ADP.Pi complexes. However, they do not seem affected by the nature of the ADP.Pi analogue (AMPPNP, ADP.BeF$_x$, S1.ADP.AlF$_4^-$ or ADP.V0$_4^{3-}$) bound to the active site. This last result is very surprising since AMPPNP and ADP.BeF$_x$ induce different S1 "lever arm" orientations to S1.ADP.AlF$_4^-$ and ADP.VO$_4^{3-}$ (Smith and Rayment 1996; Gulick et al. 1997). It would mean that the change in orientation of the "lever arm", albeit essential during the force generation process, does not modulate the structure of the interface. However, one should treat this conclusion with caution because a recent report on the 3-D structure of smooth muscle myosin S1 showed no differences between S1.ADP.BeF$_x$ and S1.ADP.AlF$_4^-$ (Dominguez et al. 1998), suggesting that the four nucleotide analogues may have the same structure.

As revealed by cross-linking reaction between more hydrophobic regions (Bertrand et al. 1997), additional contacts also take place in the A-S1.ADP.Pi state, but they should be of relatively little importance due to the overall low binding constant and the strong ionic strength dependence characterizing this interface.

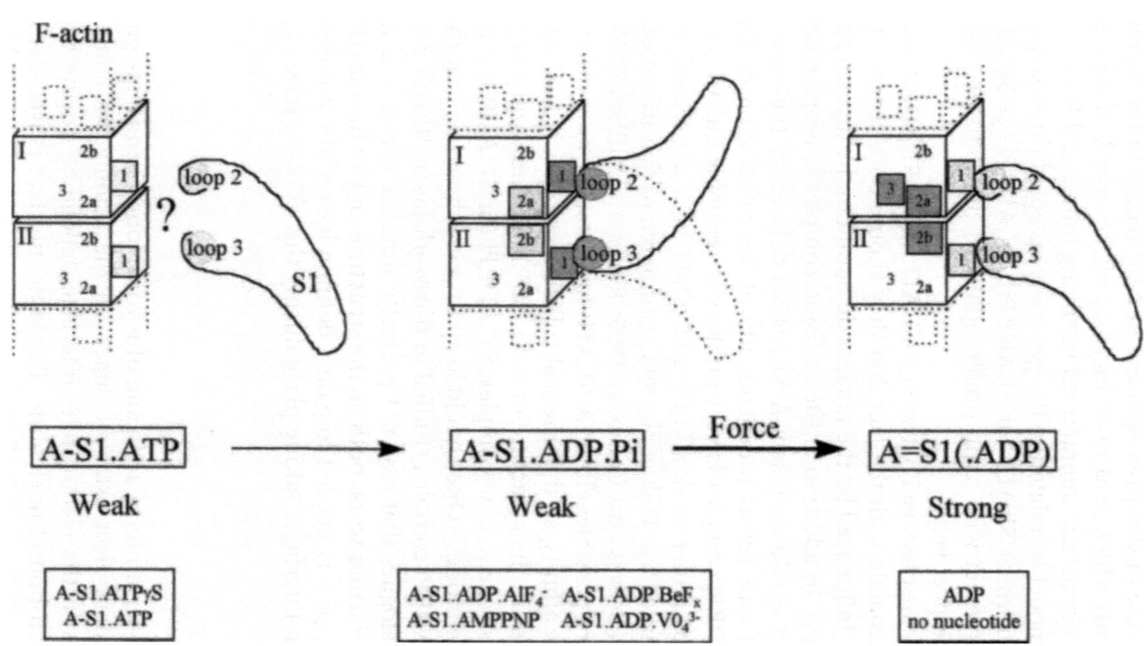

Fig. 4. Model describing the relative importance of the electrostatic contacts during actomyosin ATPase activity. Two actin monomers (*I* and *II*) of the same strand of the actin filament are depicted. The numbers of the myosin binding sites are as described in Fig. 1. The myosin motor domain (*S1*) is schematically drawn to highlight the positively charged loop 2 and loop 3 and the two extreme orientations of the C-terminal "lever arm" structure. The relative strength of the different binding sites is illustrated by *light gray* (weak) and *dark gray* (strong) shading. The ATP analogues were classified based on their effect on the cross-linking reaction performed on the actin-S1 complex

Finally, the release of the γ-phosphate and ADP results in the formation of stereospecific bonds which lead to an increase of the overall affinity and to a decrease of the electrostatic interactions involving loop 2 and loop 3.

How can this model be related to the generally accepted three-step docking model (Sect. 3) in which S1.ATP (or S1.ADP.Pi), S1′.ADP.Pi (or S1′.ADP) and S1 (or S1.ADP) interact with actin preferentially in the collision, the A-state and the R-state, respectively?

The differences we observed between A-S1.ATP and A-S1.ADP.Pi complexes are difficult to reconcile with the fact that they should both represent a collision complex not influenced by the nucleotide bound to the myosin active site (Geeves et al. 1984). In addition, if the collision complex is represented solely by A-S1.ATP, what would represent A-S1.ADP.Pi defined by the cross-linking experiments, and vice versa? Independently of the orientation of the "lever arm", the A-S1.ADP.Pi complex is very unlikely to resemble the A-state since it is characterized by rather weak affinity and rather high salt sensitivity of the actomyosin interface (Van Dijk et al. 1998), compared to that proposed for the A-state complex (Geeves and Conibear 1995; Jontes and Milligan 1996). Two interpretations are possible: either A-S1.ATP has a totally non-specific interface due to its low affinity, so that the real collision complex is A-S1.ADP.Pi, or additional changes in the structure of A-S1.ADP.Pi are necessary to generate the pre-working stroke A-state complex, S1′.ADP.Pi (or S1′.ADP, resulting in a four-state binding model). One should keep in mind, however, that the structure of the A-S1.ADP.Pi complex studied in this work (as in others) is imposed by nucleotide analogues that may not perfectly simulate the natural intermediate complexes. Further work, both on the structure and on the kinetics of the actin-myosin complex is needed to gain a definite idea of the number of different actomyosin interfaces taking place during the ATPase cycle.

7 Conclusions

Charges on the N-terminus of actin and of myosin loops 2 and 3 are good candidates for the main electrostatic contacts in the actomyosin complex. We found differences in the electrostatic interface depending on the hydrolysis state of the myosin-bound nucleotide. The ionic interface is well-defined and essential after the cleavage of the β-γ phosphate bond, compared with the non-specific contacts that seem to take place in the presence of ATP. But this ionic interface becomes of lower importance after release of both γPi and ADP in the final stereospecific complex. Altogether, these results emphasize that the structural environment of γPi in the active site alters the accessibility of the actin binding loop structures located on the surface of the protein, in agreement with the high degree of flexibility of the overall motor domain.

Note that these results and the proposed binding model are validated only for the striated muscle actomyosin complexes. The great variability of the number of charges and of the length of loop 3 could lead to a slightly different proposal with complexes of different origins; it is not unlikely that only loop 2 would then be interacting in the weak binding states. The remaining question is whether or not these contacts participate in the modulation of the enzymatic and motile properties as well as in the regulation of each myosin. Experiments using chimeric myosin with loop 2 of different myosin isoforms have already helped in tackling this question, the same type of work remains to be done with loop 3 segment.

References

Andreev OA, Borejdo J (1992) Two different acto-S1 complexes. J Muscle Res Cell Motil 13: 523–533

Andreev OA, Borejdo J (1995) Binding of heavy-chain and essential light-chain 1 of S1 to actin depends on the degree of saturation of F-actin filaments with S1. Biochemistry 34: 14829–14833

Andreev OA, Borejdo J (1997) Interaction of the heavy and light chains of cardiac myosin subfragment-1 with F-actin. Circ Res 81:688–693

Andreev OA, Andreeva AL, Borejdo J (1993) Polarization of fluorescently labeled myosin subfragment-1 fully or partially decorating muscle fibers and myofibrils. Biophys J 65:1027–1038

Arata T (1986) Structure of the actin-myosin complex produced by crosslinking in the presence of ATP. J Mol Biol 191:107–116

Bagshaw CR, Trentham DR (1974) The characterization of myosin-product complexes and of product-release steps during the magnesium ion-dependent adenosine triphosphatase reaction. Biochem J 141:331–349

Bertrand R, Chaussepied P, Kassab R, Boyer M, Roustan C, Benyamin Y (1988) Cross-linking of the skeletal myosin subfragment 1 heavy chain to the N-terminal actin segment of residues 40–113. Biochemistry 27:5728–5736

Bertrand R, Derancourt J, Kassab R (1997) Probing the hydrophobic interactions in the skeletal actomyosin subfragment 1 and its nucleotide complexes by zero-length cross-linking with a nickel-peptide chelate. Biochemistry 36:9703–9714

Blotnick E, Miller C, Grschel-Stewart U, Muhlrad D (1995) Immunochemical probing of the functional role of the 238–246 and 567–574 sequences of myosin heavy chain. Eur J Biochem 232:235–240

Bobkov AA, Bobkova EA, Lin SH, Reisler E (1996) The role of surface loops (residues 204–216 and 627–646) in the motor function of the myosin head. Proc Natl Acad Sci USA 93:2285–2289

Bonafé N, Chaussepied P. (1995) A single myosin head can be cross-linked to the N termini of two adjacent actin monomers. Biophys J 68:35s–43s

Bonafé N, Chaussepied P, Capony JP, Derancourt J, Kassab R (1993) Photochemical cross-linking of the skeletal myosin head heavy chain to actin subdomain-1 at Arg95 and Arg28. Eur J Biochem 213:1243–1254

Cantiello HF, Patenaude C, Zaner K (1991) Osmotically induced electrical signals from actin filaments Biophys J 59:1284–1289

Chaussepied P (1989) Interaction between stretch of residues 633–642 (actin binding site) and nucleotide binding site on skeletal myosin subfragment 1 heavy chain. Biochemistry 28: 9123–9128

Chaussepied P, Morales MF, Kassab R (1988) The myosin SH2–50-kiloDalton fragment cross-link: location and consequences. Biochemistry 27:1778–1785

Chen T, Applegate D, Reisler E (1985) Cross-linking of actin to myosin subfragment 1 in the presence of nucleotides. Biochemistry 24:5620–5625

Combeau C, Didry D, Carlier MF (1992) Interaction between G-actin and myosin subfragment-1 probed by covalent cross-linking. J Biol Chem 267:14038–14046

Cooke R (1997) Actomyosin interaction in striated muscle. Physiol Rev 77:671–697

Cope MJT, Whisstock J, Rayment I, Kendrick-Jones J (1996) Conservation within the myosin motor domain: implications for structure and function. Structure 4:969–987

Craig R, Greene LE, Eisenberg E (1985) Structure of the actin-myosin complex in the presence of ATP. Proc Natl Acad Sci USA 82:3247–3251

Das Gupta G, Reisler E (1992) Actomyosin interactions in the presence of ATP and the N-terminal segment of actin. Biochemistry 31:1836–1841

Diaz Banos FG, Bordas J, Lowy J, Svensson A (1996) Small segmental rearrangements in the myosin head can explain force generation in muscle. Biophys J 71:576–589

Dominguez R, Freyzon Y, Trybus KM, Cohen C (1998) Crystal structure of a vertebrate smooth muscle myosin motor domain and its complex with the essential light chain: visualization of the pre-power stroke state. Cell 94:559–571

Dos Remedios CG, Moens PD (1995) Actin and actomyosin interface: a review. Biochem Biophys Acta 1228:99–124

Duong AM, Reisler E (1989) Binding of myosin to actin in myofibrils during ATP hydrolysis. Biochemistry 28:1307–1313

Eisenberg E, Hill TL (1985) Muscle contraction and free energy transduction in biological systems. Science 227:999–1006

Fisher AJ, Smith CA, Thodenhoden J, Smith R, Sutoh K, Holden HM, Rayment I (1995) Structural studies of myosin-nucleotide complexes: a revised model for the molecular basis of muscle contraction. Biophys J 68:19s–28s

Furch M, Geeves MA, Manstein DJ (1998) Modulation of actin affinity and actomyosin ATPase by charge changes in the myosin motor domain. Biochemistry 37:6317–26

Geeves MA, Conibear PB (1995) The role of three-state docking of myosin S1 with actin in force generation. Biophys J 68:194S–199S

Geeves MA, Goody RS, Gutfreund H (1984) Kinetics of acto-S1 interaction as a guide to a model for the crossbridge cycle. J Muscle Res Cell Motil 5:351–361

Gulick AM, Bauer CB, Thoden JB, Rayment I (1997) X-ray structures of the MgADP, MgATPgammaS, and MgAMPPNP complexes of the *Dictyostelium discoideum* myosin motor domain. Biochemistry 36:11619–11628

Herrmann C, Sleep J, Chaussepied P, Travers F, Barman T (1993) A structural and kinetic study on myofibrils prevented from shortening by chemical cross-linking. Biochemistry 32: 7255–7263

Hibberd MG, Trentham DR (1986) Relationships between chemical and mechanical events during muscular contraction. Annu Rev Biophys Biophys Chem 15:119–161

Highsmith S (1990) Electrostatic contributions to the binding of myosin and myosin-MgADP to F-actin in solution. Biochemistry 29:10690–10694

Highsmith S, Murphy AJ (1992) Electrostatic changes at the actomyosin–subfragment 1 interface during force-generating reactions. Biochemistry 31:385–389

Homsher E, Lacktis J, Regnier M (1997) Strain-dependent modulation of phosphate transients in rabbit skeletal muscle fibers. Biophys J 72:1780–1791

Janin J (1997) The kinetics of protein–protein recognition. Proteins 28:153–161

Jontes JD, Milligan RA (1996) Brush border myosin-I structure and ADP-dependent conformational changes revealed by cryo-electron microscopy and image analysis. J Cell Biol 139: 683–693

Kabsch W, Mannherz HG, Suck D, Pai EF, Holmes KC (1990) Atomic structure of the actin:DNase I complex. Nature 347:37–44

Kobayashi S, Asai H, Oosawa F (1964) Electric birefringence of actin. Biochim Biophys Acta 88: 528–540

Labbé JP, Mornet D, Roseau G, Kassab R (1982) Cross-linking of F-actin to skeletal muscle myosin subfragment 1 with bis(imido esters): further evidence for the interaction of myosin head heavy chain with an actin dimer. Biochemistry 21:6897–6902

Laemmli UK (1970) Cleavage of structural proteins during the assembly of the head of bacteriophage T4. Nature 227:680–685

Lheureux K, Chaussepied P (1995a) Comparative studies of the monomeric and filamentous actin-myosin head complexes. Biochemistry 34:11435–11444

Lheureux K, Chaussepied P (1995b) Functional significance of the binding of one myosin head to two actin monomers. Biochemistry 34:11445–11452

Lheureux K, Forne T, Chaussepied P (1993) Interaction and polymerization of the G-actin-myosin head complex: effect of DNase I. Biochemistry 32:10005–10014

Lorenz M, Popp D, Holmes KC (1993) Refinement of the F-actin model against X-ray fiber diffraction data by the use of a directed mutation algorithm. J Mol Biol 234:826–836

Lymn RW, Taylor EW (1971) Mechanism of adenosine triphosphate hydrolysis by actomyosin. Biochemistry 10:4617–4624

Marianne-Pepin T, Mornet D, Bertrand R, Labbe JP, Kassab R (1985) Interaction of the heavy chain of gizzard myosin heads with skeletal F-actin. Biochemistry 24:3024–3029

Mendelson R, Morris EP (1997) The structure of the acto-myosin subfragment 1 complex: results of searches using data from electron microscopy and X-ray crystallography. Proc Natl Acad Sci USA 94:8533–8538

Miller CJ, Wong WW, Bobkova E, Rubenstein PA, Reisler E (1996) Mutational analysis of the role of the N-terminus of actin in actomyosin interactions. Comparison with other mutant actins and implications for the cross-bridge cycle. Biochemistry 35:16557–16565

Milligan RA (1996) Protein-protein interactions in the rigor actomyosin complex. Proc Natl Acad Sci USA 93:21–26

Mornet D, Bertrand R, Pantel P, Audemard E, Kassab R (1981) Structure of the actin-myosin interface. Nature 292:301–306

Onishi H, Morales MF, Katoh K, Fujiwara K (1995) The putative actin-binding role of hydrophobic residues Trp546 and Phe547 in chicken gizzard heavy meromyosin. Proc Natl Acad Sci USA 92:11965–11969

Ostap EM, Pollard TD (1996) Biochemical kinetic characterization of the *Achantamoeba* myosin-I ATPase. J Cell Biol 132:1053–1060

Patterson B, Spudich JA (1996) Cold-sensitive mutations of *Dictyostelium* myosin heavy chain highlight functional domains of the myosin motor. Genetics 143:801–810

Rayment I, Rypniewski WR, Schmidt-Base K, Smith R, Tomchick DR, Benning MM, Winkelmann DA, Wesenberg G, Holden HM (1993a) Three-dimensional structure of myosin subfragment-1: a molecular motor. Science 261:50–58

Rayment I, Holden HM, Whittaker M, Yohn CB, Lorenz M, Holmes KC, Milligan RA (1993b) Structure of the actin-myosin complex and its implications for muscle contraction. Science 261: 58–65

Roopnarine O, Thomas DD (1996) Orientation of intermediate nucleotide states of indane dione spin-labeled myosin heads in muscle fibers. Biophys J 70:2795–2806

Schreiber G, Fersht A (1996) Rapid, electrostatically assisted association of proteins. Nature Struc Biol 3:427–431

Schröder RR, Manstein DJ, Jahn W, Holden H, Rayment I, Holmes KC, Spudich JA (1993) Three-dimensional atomic model of F-actin decorated with *Dictyostelium* myosin S1. Nature 364: 171–174

Sellers JR, Goodson HV (1995) Motor proteins 2: myosin. Protein Profile 2:1323–1423

Sheterline P, Sparrow JC (1994) Actin. Protein Profile 1:1–121

Smith CA, Rayment I (1996) X-ray structure of the magnesium(II).ADP.vanadate complex of the *Dictyostelium discoideum* myosin motor domain to 1.9-Å resolution. Biochemistry 35: 5404–5417

Sutoh K (1982) Identification of myosin-binding sites on the actin sequence. Biochemistry 21: 3654–3661

Sutoh K (1983) Mapping of actin-binding sites on the heavy chain of myosin subfragment 1. Biochemistry 29:1579–1585

Tawada K, Kawai M (1990) Covalent cross-linking of single fibers from rabbit psoas increases oscillatory power. Biophys J 57:643–647

Tawada K, Kimura M (1986) Stiffness of carbodiimide-crosslinked glycerinated muscle fibres in rigor and relaxing solutions at high salt concentrations. J Muscle Res Cell Motil 7:339–350

Uyeda TQ, Ruppel KM, Spudich JA (1994) Enzymatic activities correlate with chimaeric substitutions at the actin-binding face of myosin. Nature 368:567–569

Van Dijk J, Fernandez C, Chaussepied P (1998) Effect of ATP analogues on the actin-myosin interface. Biochemistry 37:8385–8394

Xiao M, Borejdo J (1997) Orientation of cross-bridges in skeletal muscle measured with a hydrophobic probe. Biophys J 72:2268–2274

Yamamoto K (1989) ATP-induced structural change in myosin subfragment-1 revealed by the location of protease cleavage sites on the primary structure. J Mol Biol 209:703–709

The Alanine-Scanning Mutagenesis of *Dictyostelium* Myosin II at the Ionic Interface with Actin

Hideo Asukagawa and Kazuo Sutoh[1]

1 Introduction

Cyclic interaction of actin and myosin coupled with ATP hydrolysis is essential for generation of force by the actin-myosin system. The major force maintaining the actin-myosin contact during ATP hydrolysis, actin-myosin sliding and force generation seems to be ionic interactions, since these functions are highly sensitive to the ionic strength of the solvent. In the absence of ATP, i.e., under rigor conditions, however, hydrophobic interactions become dominant in maintaining the actin-myosin association. Based on the three-dimensional reconstruction of electron microscopic images of the rigor complex of actin and myosin subfragment 1 (S1), these ionic and hydrophobic interaction sites have been tentatively assigned to several locations of actin and myosin (Schroder et al. 1993; Milligan 1996).

One of these putative sites for ionic interactions resides in the so-called 50K–20K junctions of myosin. The junction can be easily identified in almost all myosins so far sequenced because it is rich in Lys, Arg and Gly residues, although their sequences are highly divergent. Since the junction sequence usually contains many Gly residues, it is expected to be highly flexible. In fact, X-ray crystallography of chicken S1 revealed that the junction was missing in the final electron density map (Rayment et al. 1993). However, three-dimensional structure of the junction of *Dictyostelium* myosin II was determined by X-ray crystallography since it was much shorter in its length, and therefore less flexible than the junctions of other myosins (Smith and Rayment 1996). The *Dictyostelium* junction has the sequence ASRAKKG (residues 618–624: DeLozanne et al. 1985), and forms a loop connecting the upper 50K and lower 50K subdomains (Fisher et al. 1995; Smith and Rayment 1996), residing at the surface interacting with actin (Fig. 1). Side chains of R620, K622 and K623 are expected point to the actin surface when myosin binds to actin. Moreover, biochemical and structural studies have shown that these basic residues are likely to be in contact with acidic residues of actin in the N-terminal segment and in a short loop in subdomain 1 containing residues 24 and 25 (Milligan 1996 and references therein). These results indicate that the ionic interactions between basic residues in the 20K–50K

[1] Department of Life Sciences, University of Tokyo at Komaba, Komaba 3-8-1, Tokyo 153-8902, Japan

Results and Problems in Cell Differentiation, Vol. 36
D.D. Thomas, C.G. dos Remedios (Eds.): Molecular Interactions of Actin

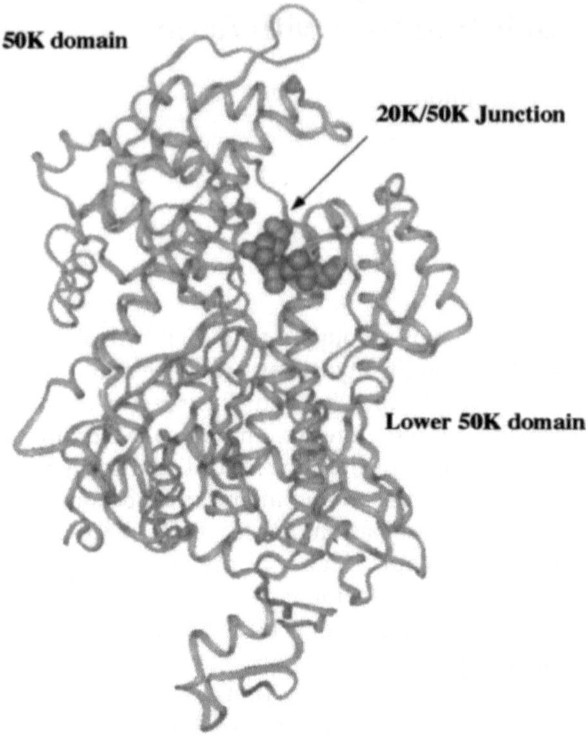

Fig. 1. Structure of the 50 K–20 K junction in the motor domain of *Dictyostelium* myosin II complexed with MgADP/ Vi. The 50 K–20 K junction with the sequence of ASRAKKG is shown in the space-filling model

junction and acidic residues on actin may play a crucial role when myosin works as a motor. In fact, it was shown that when the sequence of the 20 K–50 K junction of *Dictyostelium* myosin II was replaced with those of other myosins, actin-activated ATPase activities of the chimera myosins correlated well with those of myosins from which the junction sequences were derived (Uyeda et al. 1994). Moreover, by inserting extra basic residues with a repeating unit of the GKK motif into this 20 K–50 K junction of *Dictyostelium* myosin II, the affinity of actin and myosin in the presence of ATP was dramatically increased (Furch et al. 1998).

To investigate functional roles of the basic residues in the 20 K–50 K junction of myosin further, we carried out alanine-scanning of *Dictyostelium* myosin II, replacing a basic residue with alanine one by one (Shimada et al. 1997; Sasaki et al. 1998). Examination of in vitro and in vivo functions of the mutant myosins has shown that those basic residues in the junction really play a crucial role in maintaining ionic interactions with actin in the presence of ATP.

2 Materials and Methods

2.1 Construction and Expression of Recombinant Myosins

Each R620, K622 and K623 of *Dictyostelium* myosin II heavy chain was changed to alanine by site-directed mutagenesis (Kunkel 1985; Kunkel et al. 1987) using the *Dictyostelium* myosin II heavy chain gene (DeLozanne et al. 1985). The mutant myosin heavy chain genes were ligated to the *Dictyostelium* actin-15 promoter and *Dictyostelium* actin-6 terminator to drive their expression in *Dictyostelium* cells. They were finally inserted into a multicopy extrachromosomal vector, pBIG (Patterson and Spudich 1996). Plasmids carrying the mutant myosin heavy chain genes were introduced into *Dictyostelium* myosin-null cells in which the myosin II heavy chain gene was knocked out by means of homologous recombination (Manstein et al. 1989). *Dictyostelium* cells transformed by electroporation were selected in a medium supplemented with 20 μg/ml of G418 on a plastic dish for a week. The transformed cells thus obtained expressed the mutant myosin II molecules.

2.2 Phenotypes of the Transformed Cells

The growth rates were measured by determining the numbers of cells cultured in suspension. The incubator was shaken at 150 rpm at 22 °C. Development of the transformed cells was examined on agar plates covered with a lawn of *Escherichia coli* cells. *Dictyostelium* cells (1.2×10^4) were suspended in 10 mM TrisCl, pH 7.5, and spotted onto the bacterial lawn. When *Dictyostelium* cells cleared the bacterial cells, they had entered the developmental stage.

2.3 Myosin Purification

The following procedure was carried out at 4° C or on ice. Cells were broken by sonication in a lysis buffer comprising 10 mM Tris Cl, pH 7.5, 40 mM sodium pyrophosphate, 2 mM EDTA, 1 mM dithiothreitol, 0.3 g/ml sucrose, and protease inhibitors such as leupeptin, pepstatin A, PMSF and chymostatin. After the addition of 0.3 M KCl, the broken cells were centrifuged at 15,000 *g* for 15 min and then the supernatant was centrifuged further at 180,000 *g* for 2 h. The resulting supernatant was dialyzed against a solvent comprising 10 mM MOPS, pH 6.8, 50 mM KCl, 2 mM EDTA, 0.5 mM DTT and 0.1 mM PMSF. The actomyosin precipitate was collected by centrifugation at 15,000 *g* for 30 min. Soluble myosin was extracted from the pellet with a solvent comprising 10 mM MOPS, pH 7.4, 0.25 M NaCl, 7 mM $MgCl_2$ and 5 mM ATP. The solution was centrifuged at 560,000 *g* for 30 min. The supernatant was diluted with fivefold of a kination buffer comprising 10 mM MOPS, pH 7.4, 7 mM $MgCl_2$, 5 mM ATP and 10 µg/ml recombinant *Dictyostelium* myosin light chain kinase (MLCK; Tan and Spudich 1991). The reaction mixture was kept at room temperature for 90 min to allow the phosphorylation reaction to proceed. Then the whole solution was loaded onto an HPLC DEAE-5PW column. Myosin was eluted with 10 mM MOPS, pH 7.4, with a linear gradient of NaCl concentrations from 0 to 0.5 M. From 5 g of wet cells, approximately 300–500 µg of the phosphorylated myosin was obtained.

2.4 ATPase Assays

The actin-activated and basal MgATPase activities were measured as described previously (Itakura et al. 1993) with some modifications. A purified myosin (20–50 µg/ml) and various concentrations of F-actin (0–1 mg/ml) were mixed in an assay buffer comprising 18 mM MOPS, pH 7.4, 12.5 mM KCl, 35 mM NaCl and 5 mM $MgCl_2$. Reactions were started by adding 0.1 volume of 10 mM ATP and stopped by adding 4 volumes of 0.2 M PCA. Assays were carried out at 25 °C. The amount of released phosphate was determined with malachite green (Kodama and Kometani 1986) . The V_{max} and K_m values were calculated by plotting V/S vs S (S is the concentration of actin, and V is the rate of phosphate release).

2.5 In Vitro Motility Assays

In vitro motility assays were carried out as previously described (Kron and Spudich 1986; Toyoshima et al. 1987) with some modifications. A purified myosin (100 µg/ml) was introduced into a chamber constructed from a silicone-coated cover glass. The assay buffer comprised 25 mM imidazole, pH 7.4, 4 mM $MgCl_2$, 1 mM DTT, 0.2% of methyl cellulose, and O_2 scavengers such as catalase and glucose oxidase (Harada et al. 1990). F-actin was labeled with rhodamine/phalloidin. The sliding motion of the fluorescently labeled actin filaments was monitored under a fluorescence microscope equipped with a highly sensitive SIT camera, and recorded with a video-recorder. Assays were carried out at 25 °C.

3 Results and Discussion

3.1 Phenotypes of *Dictyostelium* Cells Expressing the Mutants

The sequence of the 50 K–20 K junction of *Dictyostelium* myosin II heavy chain is ASRAKKG (residues 618–623; Fig. 1), which is much shorter than that of skeletal muscle myosins. Each of the basic residues (Arg and Lys) in the junc-

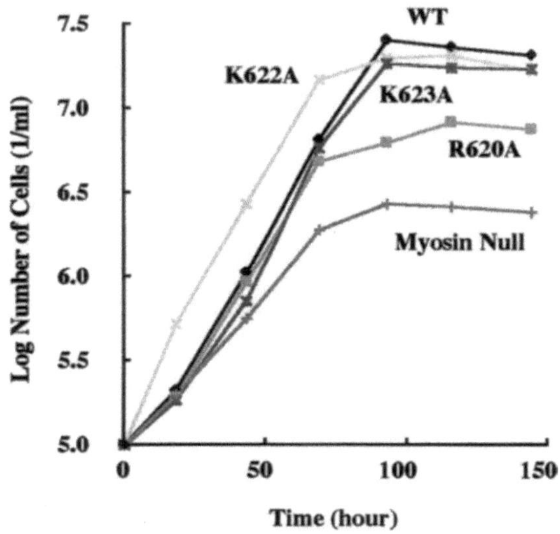

Fig. 2. Growth of the transformed cells in suspension. As controls, the growth of myosin-null cells (*Null*) and that of cells transformed with the plasmid bearing the wild type myosin heavy chain gene (*WT*) are also shown. The cells were cultured in suspension on a rotary shaker at 22 °C. The rotation rate was 150 rpm

tion was changed to Ala by introducing a site-directed mutation into the corresponding region of the *Dictyostelium* myosin heavy chain gene to generate three mutant myosins, i.e., R620 A, K622 A and K623 A. When these mutant (MT) myosins were expressed in myosin-null cells lacking the myosin II heavy chain gene (Manstein et al. 1989), they complemented defects of the myosin-null cells, and the transformants exhibited phenotypes very similar to those of the wild-type (WT) cells. Like WT cells, they grew in suspension up to a density of 2×10^7 cells/ml, although the R620 A cells saturated at a slightly lower density (Fig. 2). Moreover, they formed fruiting bodies with normal sizes upon starvation, an indication that all of these mutant myosins retained motor functions in vivo.

3.2 Characterization of Purified Mutant Myosins

The 20 K–50 K junction of *Dictyostelium* myosin forms a loop connecting the upper 50 K and lower 50 K subdomains (Fig. 1). Since side chains of the Arg and Lys residues in the loop are exposed to its surface, their replacement with Ala may not disturb the global conformation of myosin. To examine if this was the case, we measured basal Mg- and Ca-ATPase activities of WT and MT myosins (Fig. 3). MT myosins maintained similar levels of activity to WT myosin, although the former activity of MT myosins was slightly higher than

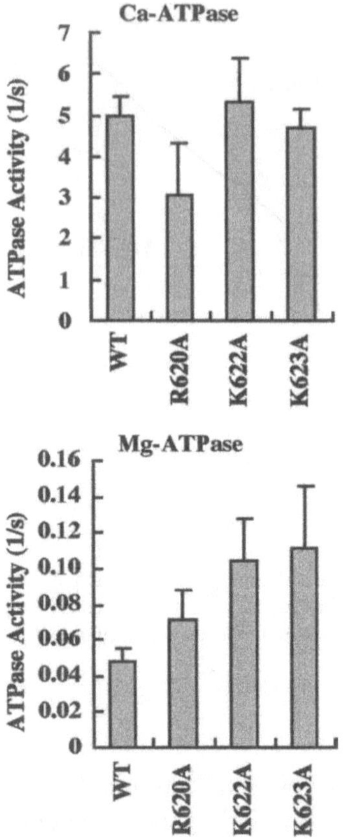

Fig. 3. Basal Mg- and Ca-ATPase activities of WT and MT myosins

that of WT myosin. The results imply that the mutations introduced into the 50K–20K junction did not induce large conformational changes in MT myosins. Unlike these ATPase activities, actin-activated ATPase activities of MT myosins were much lower than those of WT myosin at all actin concentrations examined (Fig. 4). The actin-activated ATPase activities of WT myosin reached a plateau at around 0.8 s^{-1} at actin concentrations beyond 20 μM. Those of MT myosins, however, increased almost linearly, and did not saturate even at 80 μM actin. This suggests that apparent K_m values were much larger than that of WT myosin ($K_m \approx 0.4\,\mu M$). Although it was not possible to calculate exact values from Fig. 4, K_m of MT myosins could be 50 times larger (>20 μM) than that of WT myosin, considering the fact that actin-activated ATPase activities continued to increase even at 80 μM actin. Contrary to K_m, V_{max} of MT myosins would not be significantly lower than that of WT myosin, judging from their activities at 80 μM actin.

In vitro motility assays were carried out for these mutant myosins. As shown in Fig. 5, R620 A and K622 A drove slower sliding of actin filaments, whereas

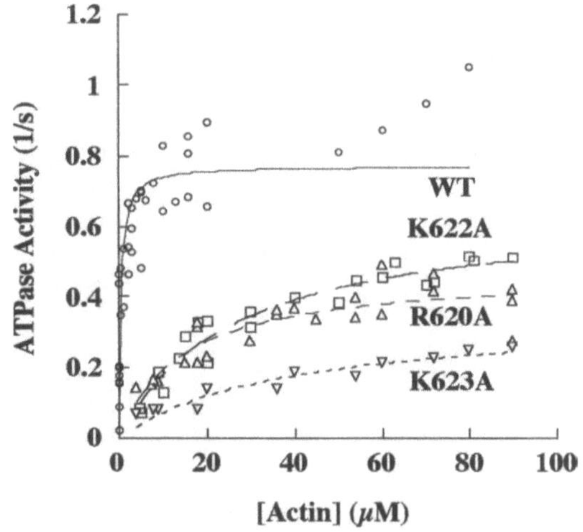

Fig. 4. Actin-activated ATPase activities of WT and MT myosins

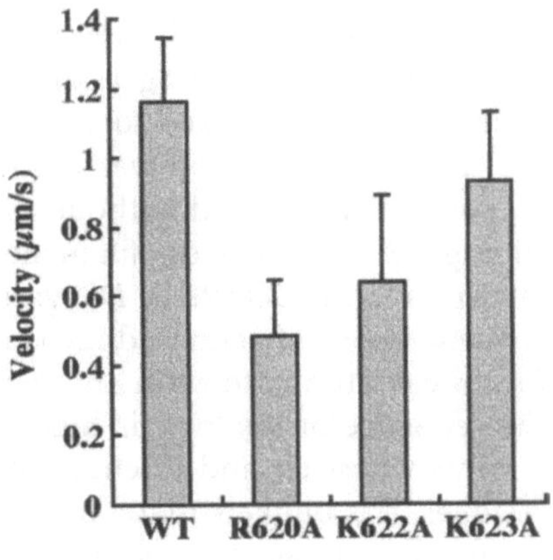

Fig. 5. In vitro motility assays of WT and MT myosins

K623 A drove the sliding at a similar velocity to WT myosin. However, when concentrations of K622 A applied on the cover glass for the motility assay were increased, the velocity of actin filaments increased gradually and reached a similar level to that of WT myosin (Fig. 6), showing that the slower sliding of actin filaments on MT myosins mainly arose from weak interaction of actin and myosin during the sliding, not from slower sliding at saturating concentrations of actin (Uyeda et al. 1990). The results are consistent with those of the actin-activated ATPase measurements mentioned above.

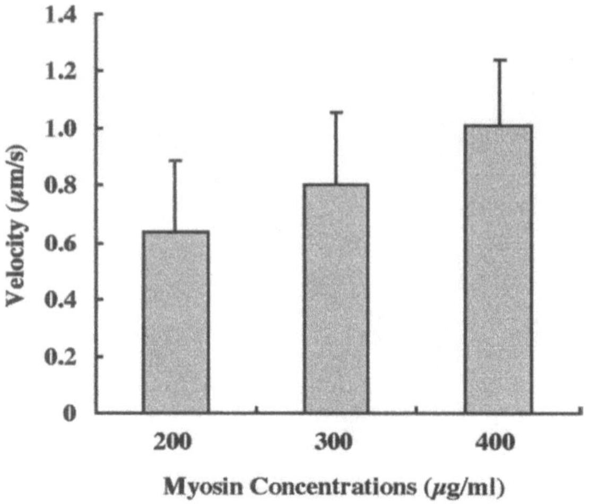

Fig. 6. Effect of myosin concentration on sliding velocity. Concentrations of K622 A loaded into the assay chamber were increased from 200 to 400 μg/ml

The present study showed the critical role of basic residues in the 20 K–50 K loop in maintaining actin-myosin association during ATP hydrolysis, actin-myosin sliding and force generation. Even when only one of the Arg or Lys residues in the loop was replaced with Ala, a remarkable increase in the K_m value was observed without significant decrease in the V_{max} value and the sliding velocity of actin filaments. Since MT myosins fully complemented myosin-specific defects in *Dictyostelium* myosin null-cells, it seems that high concentrations of actin and myosin in the cells help MT myosins to overcome their weak affinity to actin and recover full motor functions. Considering that the sequence of the loop is highly divergent among members of the myosin family (Mooseker and Cheney 1995) and that the loop is usually rich in Gly residues, it is unlikely that a particular conformation of the junction is essential for the interaction. It seems that the stable actin-myosin interaction in the presence of ATP is maintained as long as a certain number of basic residues exist in the flexible 50 K–20 K junction. These basic residues may interact with acidic residues at the N-terminus and in a short loop containing Asp24 and Asp25 in the subdomain 1 of actin to form ionic bonds as previously suggested (Sutoh et al. 1991; Johara et al. 1993).

Acknowledgements. We wish to thank Reiko Ohkura for her excellent technical assistance. The myosin II heavy chain gene, myosin-null cells, pBIG vector, and recombinant MLCK gene were provided by Dr. Spudich (Stanford University), Dr. Patterson (University of Arizona), and Dr. Uyeda (National Institute for Advanced Interdisciplinary Research, Japan). This work was supported by a Grant-in-Aid for Scientific Research from the Ministry of Education, Science and Culture of Japan, and a research grant from the Human Frontier Science Program to K.S.

References

DeLozanne A, Lewis M, Spudich JA, Leinwand LA (1985) Cloning and characterization of a non-muscle myosin heavy chain cDNA. Proc Natl Acad Sci USA 82(20):6807–6810

Fisher AJ, Smith CA, Thoden JB, Smith R, Sutoh K, Holden HM, Rayment I (1995) X-ray structures of the myosin motor domain of *Dictyostelium discoideum* complexed with MgADP.BeFx and MgADP.AlF^{4-}. Biochemistry 34(28):8960–8972

Furch M, Geeves MA, Manstein DJ (1998) Modulation of actin affinity and actomyosin adenosine triphosphatase by charge changes in the myosin motor domain. Biochemistry 37(18): 6317–6326

Harada Y, Sakurada K, Aoki T, Thomas DD, Yanagida T (1990) Mechanochemical coupling in actomyosin energy transduction studied by in vitro movement assay. J Mol Biol 216(1): 49–68

Itakura S, Yamakawa H, Toyoshima YY, Ishijima A, Kojima T, Harada Y, Yanagida T, Wakabayashi T, Sutoh K (1993) Force-generating domain of myosin motor. Biochem Biophys Res Commun 196(3):1504–1510

Johara M, Toyoshima YY, IshijimaA, Kojima H, Yanagida T, Sutoh K (1993) Charge-reversion mutagenesis of *Dictyostelium* actin to map the surface recognized by myosin during ATP-driven sliding motion. Proc Natl Acad Sci USA 90(6):2127–2131

Kodama T, Fukui K, Kometani K (1986) The initial phosphate burst in ATP hydrolysis by myosin and subfragment-1 as studied by a modified malachite green method for determination of inorganic phosphate. J Biochem (Tokyo) 99(5):1465–1472

Kron SJ, Spudich JA (1986) Fluorescent actin filaments move on myosin fixed to a glass surface. Proc Natl Acad Sci USA 83(17):6272–6276

Kunkel TA (1985) Rapid and efficient site-specific mutagenesis without phenotypic selection. Proc Natl Acad Sci USA 82(2):488–492

Kunkel TA, Roberts JD, Zakour RA (1987) Rapid and efficient site-specific mutagenesis without phenotypic selection. Methods Enzymol 154(367):367–382

Manstein DJ, Titus MA, De LA, Spudich JA (1989) Gene replacement in *Dictyostelium*: generation of myosin null mutants. EMBO J 8(3):923–932

Milligan RA (1996) Protein–protein interactions in the rigor actomyosin complex. Proc Natl Acad Sci USA 93(1):21–26

Mooseker MS, Cheney RE (1995) Unconventional myosins. Annu Rev Cell Dev Biol 11:633–675

Patterson B, Spudich JA (1996) Cold-sensitive mutations of *Dictyostelium* myosin heavy chain highlight functional domains of the myosin motor. Genetics 143(2):801–810

Rayment I, Rypniewski WR, Schmidt BK, Smith R, Tomchick DR, Benning MM, Winkelmann DA, Wesenberg G, Holden HM (1993) Structure of the actin-myosin complex and its implications for muscle contraction. Science 261(5117):50–58

Sasaki N, Shimada T, Sutoh K (1998) Mutational analysis of the switch II loop of *Dictyostelium* myosin II. J Biol Chem 273(32):20334–20340

Schroder RR, Manstein DJ, Jahn W, Holden H, Rayment I, Holmes KC, Spudich JA (1993) Three-dimensional atomic model of F-actin decorated with *Dictyostelium* myosin S1. Nature 364 (6433):171–174

Shimada T, Sasaki N, Ohkura R, Sutoh K (1997) Alanine scanning mutagenesis of the switch I region in the ATPase site of *Dictyostelium discoideum* myosin II. Biochemistry 36(46): 14037–14043

Smith CA, Rayment I (1996) X-ray structure of the magnesium(II).ADP.vanadate complex of the *Dictyostelium discoideum* myosin motor domain to 1.9 Å resolution. Biochemistry 35(17): 5404–5417

Sutoh K, Ando M, Sutoh K, Toyoshima YY (1991) Site-directed mutations of *Dictyostelium* actin: disruption of a negative charge cluster at the N terminus. Proc Natl Acad Sci USA 88(17): 7711–7714

Tan JL, Spudich JA (1991) Characterization and bacterial expression of the *Dictyostelium* myosin light chain kinase cDNA. Identification of an autoinhibitory domain. J Biol Chem 266(24): 16044–16049

Toyoshima YY, Kron SJ, McNally EM, Niebling KR, Toyoshima C, Spudich JA (1987) Myosin subfragment-1 is sufficient to move actin filaments in vitro. Nature 328(6130):536–539

Uyeda TQ, Kron SJ, Spudich JA (1990) Myosin step size. Estimation from slow sliding movement of actin over low densities of heavy meromyosin. J Mol Biol 214(3):699–710

Uyeda TQ, Ruppel KM, Spudich JA (1994) Enzymatic activities correlate with chimaeric substitutions at the actin-binding face of myosin. Nature 368(6471):567–569

Familial Hypertrophic Cardiomyopathic Myosin Mutations That Affect the Actin-Myosin Interaction

Osha Roopnarine[1]

1 Description of the FHC Disease

Familial hypertrophic cardiomyopathy (FHC) is primarily an autosomal dominant human cardiac disease, but has shown sporadic tendencies. It is characterized by left ventricular hypertrophy and myocellular disarray (Davies 1984; Maron et al. 1987). The disease is sometimes associated with skeletal myopathy of slow muscle fibers. The disease is clinically variable, ranging from benign to severe and is often the cause of sudden death in young athletes (Maron et al. 1978, 1986). The variability in the phenotype of the disease makes it a difficult disease to diagnose. Genetic linkage analyses have shown that FHC is also genetically heterogeneous because it is caused by multiple mutations in at least eight different genes (Solomon et al. 1990; Dausse and Schwartz 1993; Watkins et al. 1995c). All the mutations that cause FHC are found in sarcomeric cardiac proteins; β-myosin heavy chain (Geisterfer-Lowrence et al. 1990; Watkins et al. 1993), α-tropomyosin (Watkins et al. 1995b), troponin T (Thierfelder et al. 1994), myosin binding protein-C (Watkins et al. 1995a; Bonne et al. 1995), ventricular essential light chain (ELC), ventricular regulatory light chain (RLC; Poetter et al. 1996; Flavigny et al. 1998), troponin I (Kimura et al. 1997), and actin (Mogensen et al. 1999).

This review focuses on the myosin heavy chain (HC) and LC FHC mutations, with a brief mention of actin FHC mutations. The relationship between the functional properties of a few myosin mutations and phenotype (benign, intermediate, or severe) of the disease will be discussed. Cardiovascular research is poised to move beyond the discovery that mutations in muscle proteins cause FHC. It is critically important to determine the molecular mechanisms that are involved in the disease and to determine its impact on current models for muscle contraction, in which it is becoming increasing clear that structural transitions within myosin play a crucial role. This will lead to a better understanding of this human cardiac disease and will assist in developing therapeutic methods for managing the disease.

[1] Department of Biochemistry, 6-155 Jackson Hall, 312 Church St. SE, University of Minnesota Medical School, Minneapolis, Minnesota 55455, USA

Results and Problems in Cell Differentiation, Vol. 36
D.D. Thomas, C.G. dos Remedios (Eds.): Molecular Interactions of Actin

2 Myosin FHC Mutations

At least 45 different myosin HC mutations have been found to cause the FHC disease (reviewed by Vikstrom and Leinwand 1996). These mutations account for about 30–35% of all the reported FHC cases (Watkins et al. 1995c). Most of the mutations appear in clusters near functionally significant regions of the myosin motor domain (S1) such as the actin-binding region, the ATP catalytic site, and the LC-binding domain (Fig. 1; summarized in Table 1; Rayment et al. 1995). A number of mutations have also been found at the junction between S1 and S2 (Rayment et al. 1995). It is clear that specific amino acid residues play crucial roles in the ATPase function of the myosin molecule

Table 1. Myosin FHC mutations relative to the functional domains. Phenotype of FHC mutations: bold indicates severe phenotype; italics, intermediate; underlined, benign; plain, unclear phenotype

Functional domains in myosin	FHC mutation
ATP binding domain	Tyr124Ile, Tyr162Cys, Asn187Lys, Gln222Lys, <u>Asn232Ser</u>, *Arg249Gln*, <u>Gly256Glu</u>, **Arg453Cys**
Actin interface	**Arg403Gln, Arg403Leu**, <u>Arg403Trp</u>, <u>Val606Met</u>, Arg663His
Converter region (SH-SH2 region)	<u>Phe513Cys</u>, *Gly584Arg*
LC domain interface	Gly716Arg, <u>Arg719Gln</u>, **Arg719Trp**, Arg723Cys, Gly741Arg, **Asp778Gly**, Ser782Asn
ELC	Met149Val, Arg154His
RLC	Ala13Thr, Phe18Leu, Glu22Lys, Arg58Gln, Pro94Ala

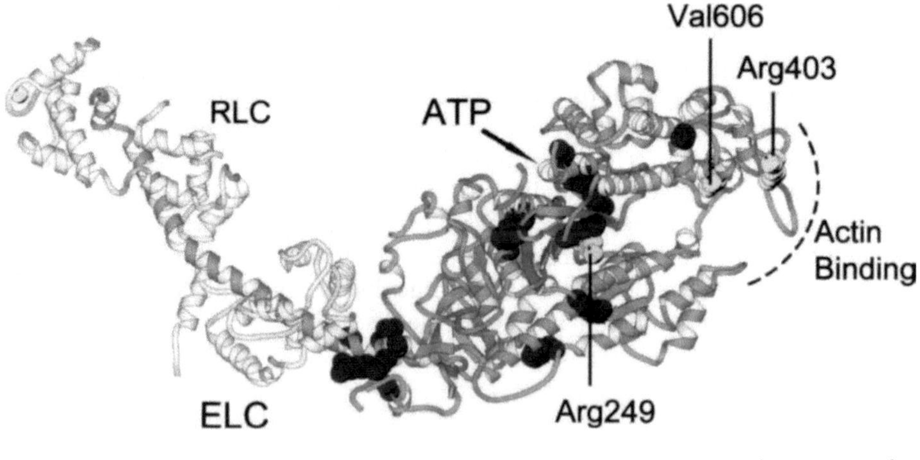

Fig. 1. The crystal structure of chicken skeletal S1 showing 21 FHC mutations (Rayment et al. 1995) relative to the actin-, ATP-, and LC-binding region of myosin. The figure (Fig. 1 in Roopnarine and Leinwand 1998) was prepared using the program MOLSCRIPT, version 2.0. (Kraulis 1991)

(reviewed in Rayment et al. 1995). Reconstruction of crystal structures of the rigor actin-myosin S1 complex have led to the proposal that specific residues, loops, α-helices, and β-sheets may be directly involved in the mechanism for force transduction (reviewed by Milligan 1996; Vale 1996). The proximity of the FHC mutations to the functional regions suggests that the motor function of the mutant myosin may be impaired in diseased patients.

3 Myosin FHC Mutations That Affect the Actin-Myosin Interaction

Progress in correlating the functional impairment due to the myosin-FHC mutation and the phenotype of the disease has been impeded by several factors: (1) difficulty in obtaining diseased human tissue, (2) in most cases when diseased human tissue is obtained, it is almost impossible to distinguish the mutant myosin from normal myosin, (3) expression of sufficient mutant myosins in an in vitro system, and (4) creating a mouse model for the disease is expensive and time-consuming.

3.1 Myosin FHC Mutations Near the Actin-Binding Interface

The Arg403Gln mutation is associated with a high frequency of sudden death in young adults and heart failure of adults in their fourth decade (Epstein et al. 1992; Watkins et al. 1992; Marian et al. 1995a). Arg403 (the equivalent residue in chicken skeletal S1 is cArg405) is at the base of a loop (cArg405-cLys415), which lies close to the actin residues, Pro332-Glu334, in the reconstruction of rigor actin-S1 complex and may play an important role during the stereospecific actin-myosin interaction (Fig. 2; Rayment et al. 1993; Milligan 1996). Reconstructions of the actin-S1 crystal structures led to the proposal that the early weak-binding interaction between myosin and actin involves charged residues, while the later strong-binding interaction involves hydrophobic interactions (Rayment et al. 1993; Schroder et al. 1993). The Arg403Gln mutation decreases the positive charge by one and may be likely to weaken the actomyosin interaction during force generation. Arg403 is a "hot spot" for FHC mutations as two other independent mutations of this residue that, cause FHC have been found: Arg403Leu and Arg403Trp (Dausse et al. 1993; Posen et al. 1995). The Arg403Leu mutation results in variable phenotype in that, diseased patients show either a high incidence of sudden death in young adults or survival into and past the sixth decade (Dausse et al. 1993). While the Arg403Trp mutation results in a benign phenotype and shows long-

The corresponding residues in the crystal structure of chicken skeletal S1 is preceded by "c", for example, cArg251.

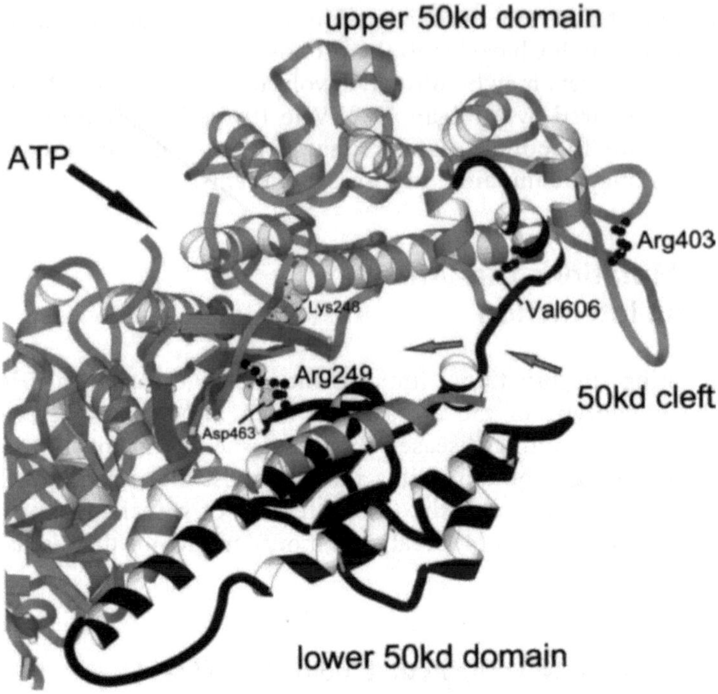

Fig. 2. An expanded region of the myosin head showing the upper 50-kDa domain, lower 50-kDa domain, and the 50-kDa cleft and three FHC mutations (Arg403Gln, Val606Met, and Arg249Gln). The figure (Fig. 5 in Roopnarine and Leinwand 1998) was prepared using the program MOLSCRIPT, version 2.0. (Kraulis 1991)

term survival of patients (Dausse et al. 1993; Posen et al. 1995), suggesting that a change of charge is not solely responsible for the clinical features of the disease, so this property of the mutation is not a good indicator for the severity of FHC disease.

The β-myosin is also expressed in slow skeletal muscle and the mutant myosins (Arg403Gln and Leu908Val) are expressed at both the mRNA and protein level (Cuda et al. 1993). Soleus muscle biopsies of diseased patients have revealed abnormalities resembling central core disease, which is a non-progressive skeletal myopathy associated with the loss of mitochondria from the center of the muscle fiber (Fananapazir et al. 1993). Total myosin (mixture of mutant- and normal-β myosin) isolated from either cardiac or soleus muscle biopsies from patients with the Arg403Gln mutation decreased the in vitro actin filament motility by 82% (Cuda et al. 1993; Cuda et al. 1997a). Analysis of the isolated myosin for some of these patients showed that the mutant was present in ~50%, reflecting the dominant negative property of the mutation. Single muscle fiber mechanics of soleus muscle from other diseased patients with the Arg403Gln mutation showed both decreased force/stiffness ratio (56% of normal fibers) and velocity of shortening (50% of normal; Lankford et al. 1995). In contrast, the isometric tension of single fibers from muscle biopsies of diseased patients with the Arg403Gln mutation was decreased by 18% compared with tension from normal patients (Malinchik et al. 1997), even though ~50% of the myosin contained the Arg403Gln mutation. It is possible that other factors (not included in the report) play a role in the manifestation of the disease, such as stress, age, and exercise.

The actin-activated myosin ATPases of expressed Arg403Gln-myosin fragments created either in the background of human cardiac β-myosin (Sata and Ikebe 1996), rat cardiac α-myosin (Sweeney et al. 1994; Roopnarine and Leinwand 1998), or *Dictyostelium* myosin (Fujita et al. 1997) showed impairment of ATPase activity that correlates with the phenotype of the disease. These studies show that the V_{max} of Arg403Gln-myosin fragments (either HMM or S1) decreased by 3.5-fold and the K_m increased by 3.4-fold compared with normal S1, suggesting that the actin-myosin interaction is weakened substantially. Therefore, the severe impairment of myosin functional properties by the Arg403Gln mutation correlates with the poor prognosis displayed by diseased patients.

The phenotype of the Val606Met mutation is variable in the extreme, it was initially reported to be a benign form of the disease in which most patients survived beyond the sixth decade (Watkins et al. 1992, Marian et al. 1995a). However, the Val606Met mutation results in a severe phenotype in another study with an unrelated family in which 50% of the diseased patients died between the ages of 15 and 27 years, demonstrating the variability of FHC (Fananapazir and Epstein 1994). This complicates the interpretation of results from functional studies. Val606 is in the actin-binding domain and appears to be buried in the upper 50-kDa domain of the myosin head. Force measurements of soleus muscles from a patient with this mutation were normal in one report (Thedinga et al. 1996). However, myosin isolated from soleus muscle from a diseased patient in another family showed that the mutant myosin reduced the in vitro actin filament sliding velocity by 2.3-fold (Cuda et al. 1997a). The V_{max} and K_m of actin-activated myosin ATPases (HMM- or S1-Val606Met) were either unaffected or affected to a lesser extent compared with the wild-type protein, respectively (Sata and Ikebe 1996; Roopnarine and Leinwand 1998). The V_{max} decreased by 1.2-fold and the K_m increased by 1.4-fold compared with wild-type S1, suggesting a slightly decreased actomyosin interaction (Roopnarine and Leinwand 1998). These differences suggest that a small perturbation in myosin biochemical function may result in a variable prognosis for patients. It is possible that this difference is due to differences in the amount of mutant protein in the muscle fiber or cardiac myocyte since this has not been determined for the Arg403Gln mutation. It is also likely that other factors such as genetic, environmental, and/or exercise influences the phenotype and onset of the FHC disease. The mild effect of the Val606Met mutation on the functional property of myosin correlates well with the benign phenotype for diseased patients.

3.2 Myosin FHC Mutations Near the ATP Binding Site

The Arg249Gln mutation has an intermediate phenotype in diseased patients (Watkins et al, 1992). Arg249 is located at the base of the ATP binding pocket about 29 Å from the phosphate-binding loop (Fig. 2). It is exposed to the surface of the molecule as it faces toward the cleft that splits the 50-kDa domain of S1 (Rayment et al. 1995). It is proposed that the movement of the lower domain relative to the upper domain is directly coupled to a structural change in the LC domain region of myosin that is responsible for the power stroke during force generation (Fisher et al. 1995). It is possible that the Arg249Gln mutation may perturb movements in the lower 50-kDa domain (highlighted in black in Fig. 2) that involves closure of the cleft during transition from the prehydrolysis state (myosin.ATP) to a metastable state (myosin.ADP.Pi; Fisher et al. 1995). Alternatively (or additionally), the proximity of Arg249 to residues that are directly involved in the mechanism of ATP binding and hydrolysis via charge-dependent interactions suggests that any charge changes in Arg249 may disrupt the myosin ATPase reaction. Arg249 (cArg251) is 11.6 Å from cAsp463, which forms a hydrogen bond with cThr186 (cThr186 is involved in the coordination of the Mg^{2+} ion of the bound ATP; Smith and Rayment 1996). Arg249 is close to cLys248, which is proposed to be directly involved in a salt bridge with cAsp463 during ATP binding (Fig. 2; Smith and Rayment 1996).

The V_{max} of Arg249Gln-S1 decreased by 1.7-fold and the K_m increased by 1.85-fold compared with normal S1, suggesting that an actin-dependent kinetic rate constant of the ATPase cycle may be affected (Roopnarine and Leinwand 1998). Similar results were reported with Arg249Gln-HMM (Sata and Ikebe 1996), however, they reported that K_m was unchanged but the actin motility decreased by 1.6-fold (Sata and Ikebe 1996). The functional properties of myosin with Arg249Gln correlates well with the moderate phenotype associated with the diseased patients.

4 Light Chain FHC Mutations That Affect the Actin-Myosin Interaction

The FHC mutations in human ventricular ELC are Met149Val and Arg154His. The affected patients had classical FHC phenotypes as well as mid-left ventricular chamber (MVC) thickening, and ragged red fiber (RRF) disease of the skeletal muscle (deltoid and soleus muscle). The N-terminal residues (1–41) of ELC have been strongly implicated in binding to actin and play a modulatory role in the actin-activated ATPase rate of myosin (Timson et al. 1998).

While the C-terminal 14 residues and the central domain of ELC appear to bind to the heavy chain of myosin and may be involved in a stabilizing role (Rayment et al. 1993), which may be perturbed by the ELC FHC mutations (Arg149 and Met 154). The Met149Val ELC mutation in myosin induce slightly elevated in vitro actin filament motility, even though the affected patients have clinical features of the disease in their heart and skeletal muscles (Poetter et al. 1996).

The FHC mutations in RLC are Ala13Thr, Glu22Lys, Phe18Leu, Arg58Gln, and Pro94Ala (Poetter et al. 1996; Flavigny et al. 1998). In addition to having the classical disease phenotype, two of the mutations (Ala13Thr and Glu22Lys) resulted in MVC thickening in the affected patients. The Glu22Lys and Pro94Ala mutations are in the interface between the N-terminal and C-terminal domains of RLC. The C-terminal and central regions of RLC are also proposed to interact with the heavy chain (Rayment et al. 1993). It was suggested that residues 71–91 may interact with the heavy chain (Wadgaonkar et al. 1993), which may be perturbed by the Pro94Ala mutation in RLC. The N-terminus of RLC contains the highly phosphorylatable Ser15, Ser19 and Thr24, which are believed to play a regulatory role in muscle contraction. The RLC FHC mutations may perturb the charge of the N-terminus.

Myosin containing Glu22Lys RLC that was isolated from human muscle biopsies induced normal actin filament motility (Poetter et al. 1996). However, deltoid muscle fiber biopsies from patients containing the RLC mutation, Glu22Lys, show increased calcium sensitivity during force production and increased relaxed disorder compared with that from unaffected people (Levine et al. 1998). The mean force/cross-sectional area of muscle biopsies from patients with the Glu22Lys was similar to fiber samples from unaffected people, however the mutant fibers had almost twice the cross-sectional area compared with normal fibers (Levine et al. 1998).

5 Actin Mutations That Cause FHC

Actin is a highly conserved protein, so it is not surprising that the incidence of a disease-causing mutation is very low. To date, only one mutation in the actin gene (Ala295Ser) has been found to cause FHC (Mogensen et al. 1999). Ala295 is located in domain 3 of actin on the surface of a pocket that has both polar and hydrophobic residues, whose interaction may be affected by the introduction of a polar side group on serine. The FHC-diseased family that was investigated had variable phenotypes, but the majority of the diseased individuals had no symptoms. Two other actin mutations (Arg312His and Glu361Gly) were found to cause heart disease, idiopathic dilated cardiomyopathy (Olson et al. 1998). The Arg312His mutation is in domain 3, away from the region in actin that is proposed to directly interact with myosin. However, Glu361Gly is in domain 1 of actin, which is the key actin domain involved with interaction with myosin (Rayment et al. 1993). Reconstruction of the actin-S1 crystal complex has led to the proposal of specific sites on the actin-S1 interface that are involved in weak- and strong-binding interactions (Rayment et al. 1993; Milligan 1996). Several biochemical and structural studies have supported these proposals as reviewed in this book.

6 Mechanism for the FHC Disease

It is becoming clear that a defect in myosin function due to the presence of an FHC mutation correlates with the severity of the disease. The fact that myosin mutations in different functional locations in the head result in impaired actin interactions suggests that the functional properties of myosin (ATPase activity, actin binding and translocation, and light chain domain movements) are coupled together. So, despite the fact that there are specific myosin residues that are proposed to interact directly with specific actin residues at the actin-myosin interface, residues distal from this region appear to perturb the actin-myosin interaction. The varying locations and residue replacements that cause the FHC disease suggest that a single mechanism is not responsible for causing the disease.

Protein analysis of muscle biopsies from diseased patients revealed that myosin containing the Arg403Gln mutation is expressed almost equally compared with normal myosin, but the functional properties of the mutant myosin is more than 50% impaired, supporting the dominant negative characteristic of the disease (Cuda et al. 1993; Malinchick et al. 1997). It was shown that mixing a slower myosin (i.e., one with a lower ATPase rate) with a fast myosin (i.e., one with a higher ATPase rate) results in a dominant negative effect on the velocity of sliding of actin filaments (Cuda et al. 1997a,b).

The mechanism for muscle cell hypertrophy and myocyte disarray by FHC mutations is still unknown. Results from functional studies of the mutant proteins make it clear that mutations in sarcomeric proteins are likely to compromise sarcomere function (either in terms of motor function, force production, and/or protein structure) resulting in hypertrophic cardiomyopathy. Cardiac hypertrophy can be the result of a number of stimuli, including exercise, pressure, and volume overload. It is possible that the mutant sarcomeric protein perturbs myocardial contractility of the myocyte, which imposes stress on the myocytes to induce a hypertrophic response. It is apparent that the hypertrophic response to the FHC mutation is more detrimental to the cardiac muscle than the slow-twitch skeletal muscle fiber that displays hypertrophy due to the same mutation. The introduction of human β Arg403Gln-myosin into feline cardiac myocytes disrupts the sarcomeric ultrastructure, which may be due to an increased protein turnover rate or impaired mutant myosin binding to actin (Marian et al. 1995b). The latter agrees with the impaired acto-myosin interaction of the Arg403Gln-myosin fragments, suggesting that the biochemical defects of the motor domain may disturb myosin filament assembly in the sarcomere. However, in another report, Arg249Gln- and Arg403Gln-myosins assembled normally into the thick filaments of rat cardiac myocytes, with no detectable disruption of myofilament assembly (Becker et al. 1997). The differences between these two reports (Marian et al. 1995b; Becker et al. 1997) may be due to differences in the cardiac myocytes (feline vs rat) and/or the amount of mutant protein incorporated in a sarcomere (different techniques we used to introduce the mutant myosins).

7 Conclusions

The results from functional studies suggest that the degree of disruption of the actomyosin interaction by the myosin FHC mutations may be directly correlated with the severity of the disease. It is clear that both the location of the FHC mutation relative to the functional regions of the protein and the chemical nature of the FHC mutation affect the clinical outcome of diseased patients. It is anticipated that a correlation between the functional defects and the phenotype of the disease will influence general therapeutic ways for managing it. Therefore, biochemical analysis of several other FHC mutations will better establish a definite correlation between functional impairment and the clinical phenotype of this heart disease.

References

Becker KD, Gottshall KR, Hickey R, Perriard JC, Chien K (1997) Point mutations in human beta cardiac myosin heavy chain have differential effects on sarcomeric structure and assembly: an ATP binding site change disrupts both thick and thin filaments, whereas hypertrophic cardiomyopathy mutations display normal assembly. J Cell Biol 137:131–140

Bonne G, Carrier L, Bercovici J, Cruaud C, Richard P, Hainque B, Gautel M, Labeit S, James M, Weissenbach J, Vosberg H, Fiszman M, Komajda M, Schwartz K (1995) A splice acceptor site mutation in the cardiac myosin binding protein C gene is associated with familial hypertrophic cardiomyopathy. Nat Genet 11:438–440

Cuda G, Fananapazir L, Zhu WS, Sellers JR, Epstein ND (1993) Skeletal muscle expression and abnormal function of beta-myosin in hypertrophic cardiomyopathy. J Clin Invest 91:2861–2865

Cuda G, Fananapazir L, Epstein ND, Sellers JR (1997a) In vitro motility activity of β-cardiac myosin depends on the nature of the β-myosin heavy chain gene mutation in hypertrophic cardiomyopathy. J Mus Res Cell Motil 18:275–283

Cuda G, Pate E, Cooke R, Sellers JR (1997b) In vitro actin filament sliding velocities produced by mixtures of different types of myosin. Biophys J 72:1767–1779

Dausse E, Schwartz K (1993) Genetic heterogeneity of familial hypertrophic cardiomyopathy. Neuromusc Disord 3:483–486

Dausse E, Komajda M, Fetler L, Dubourg O, Dufour C, Carrier L, Wisnewsky C, Bercovic J, Hengstenberg C, al-Mahdawi S, Isnard R, Hagege A, Bouhour J, Desnos M, Beckman J, Wessenbach J, Schwartz K, Guicheney PJ (1993) Familial hypertrophic cardiomyopathy. Microsatellite haplotyping and identification of a hot spot for mutations in the beta-myosin heavy chain gene. J Clin Invest 92:2807–2813

Davies MJ (1984) The current status of myocardial disarray in hypertrophic cardiomyopathy. Br Heart J 51:361–366

Epstein ND, Cohn GM, Cyran F, Fananapazir L (1992) Differences in clinical expression of hypertrophic cardiomyopathy associated with two distinct mutations in the beta-myosin heavy chain gene. A 908LeuVal mutation and a 403ArgGln mutation. Circ 86:345–352

Fananapazir L, Epstein ND (1994) Genotype-phenotype correlations in hypertrophic cardiomyopathy. Insights provided by comparisons of kindreds with distinct and identical beta-myosin heavy chain gene mutations. Circ 89:22–32

Fananapazir L, Dalakas MC, Cyran F, Cohn G, Epstein ND (1993) Missense mutations in the beta-myosin heavy-chain gene cause central core disease in hypertrophic cardiomyopathy. Proc Natl Acad Sci USA 90:3993–3997

Fisher AJ, Smith CA, Thoden J, Smith R, Sutoh K, Holden HM, Rayment I (1995) X-ray structures of the myosin motor domain of *Dictyostelium discoidium* complexed with MgADP.BeF$_x$ and MgADP.ALF$_4^-$. Biochemistry 34:8960–8972

Flavigny J, Richard P, Isnard R, Carrier L, Charron P, Bonne G, Forissier JF, Desnos M, Dubourg O, Komajda M, Schwartz K, Hainque B (1998) Identification of two novel mutations in the ventricular regulatory myosin light chain gene (MYL2) associated with familial and classical forms of hypertrophic cardiomyopathy. J Mol Med 76:208–214

Fujita H, Sugiura S, Momomura S, Omata M, Sugi H, Sutoh K (1997) Characterization of mutant myosins of *Dictyostelium discoideum* equivalent to human familial hypertrophic cardiomyopathy mutants. J Clin Invest 99:1010–1015

Geisterfer-Lowrance AA, Kass S, Tanigawa G, Vosbert HP, McKenna W, Seidman JG, Seidman CE (1990) A molecular basis for familial hypertrophic cardiomyopathy: A β cardiac myosin heavy chain missense mutation. Cell 62:999–1006

Kimura A, Harada H, Park JE, Nishi H, Satoh M, Takahashi M, Hiroi S, Sasaoka T, Ohbuchi N, Nakamura T, Koyanagi T, Hwang TH, Choo JA, Chung KS, Hasegawa A, Nagai R, Okazaki O, Nakamura H, Matsuzaki M, Sakamoto T, Toshima H, Koga Y, Imaizumi T, Sasazuki T (1997) Mutations in the cardiac troponin I gene associated with hypertrophic cardiomyopathy. Nat Genet 16:379–382

Kraulis PJ (1991) MOLSCRIPT: A program to produce both detailed and schematic plots of protein structures. J Applied Crystall 24:946–950

Lankford EB, Epstein ND, Fananapazir L, Sweeney HL (1995) Abnormal contractile properties of muscle fibers expressing beta-myosin heavy chain gene mutations in patients with hypertrophic cardiomyopathy. J Clin Invest 95:1409–1414

Levine RJ, Yang Z, Epstein ND, Fananapazir L, Stull JT, Sweeney HL (1998) Structural and functional responses of mammalian thick filaments to alterations in myosin regulatory light chains. J Struct Biol 122:49–161

Malinchik S, Cuda G, Podolsky RJ, Horowits R (1997) Isometric tension and mutant myosin heavy chain content in single skeletal myofibers from hypertrophic cardiomyopathy patients.J Mol Cell Cardiol 1997 29:667–76

Marian AJ, Mares A, Kelly DP, Yu QT, Abchee AB, Hill R, Roberts R (1995a) Sudden death in hypertrophic cardiomyopathy. Variability in phenotype expression of β-myosin heavy chain mutations. Eur Heart J 16:368–376

Marian AJ, Yu QT, Mann DL, Graham FL, Roberts R (1995b) Expression of a mutation causing hypertrophic cardiomyopathy disrupts sarcomere assembly in adult feline cardiac myocytes. Circ Res 77:98–106

Maron BJ, Roberts WC, Edwards JE, McAllister HA, Foley DD, Epstein SE (1978) Sudden death in patients with hypertrophic cardiomyopathy. Characterization of 26 patients without functional limitation. Am J Cardiol 41:803–810

Maron BJ, Roberts WC, Epstein SE (1986) Causes of sudden death in competitive athletes. J Am Coll Cardiol 7:204–214

Maron BJ, Bonow RO, Cannon RO, Leon MB, Epstein SE (1987) Hypertrophic cardiomyopathy: interrelations of clinical manifestations, pathophysiology and therapy. N Engl J Med 316: 780–789

Milligan RA (1996) Protein–protein interactions in the rigor actomyosin complex. Proc Natl Acad Sci USA 93:21–22

Mogensen J, Klausen IC, Pedersen AK, Egeblad H, Bross P, Kruse TA, Gregersen N, Hansen PS, Baandrup U, Borglum AD (1999) Alpha-cardiac actin is a novel disease gene in familial hypertrophic cardiomyopathy. J Clin Invest 103:R39–43

Olson TM, Michels VV, Thibodeau SN, Tai YS, Keating MT (1998) Actin mutations in dilated cardiomyopathy, a heritable form of heart failure. Science 280:750–752

Poetter K, Jiang H, Hassanzadeh S, Master SR, Chang A, Dalakas MC, Rayment I, Sellers JR, Fananapazir L, Epstein ND (1996) Mutations in either the essential or regulatory light chains of myosin are associated with a rare myopathy in human heart and skeletal muscle. Nat Genet 13:63–69

Posen BM, Moolman JC, Corfield VA, Brink PA. (1995) Clinical and prognostic evaluation of familial hypertrophic cardiomyopathy in two South African families with different cardiac β myosin heavy chain gene mutations. Br Heart J 74:40–46

Rayment I, Holden HM, Whittaker M, Yohn CB, Lorenz M, Holmes KC, Milligan RA (1993) Structure of the actin-myosin complex and its implications for muscle contraction. Science 261: 58–65

Rayment I, Holden HM, Sellers JR, Fananapazir L, Epstein ND (1995) Structural interpretation of the mutations in β-cardiac myosin that have been implicated in familial hypertrophic cardiomyopathy. Proc Natl Acad Sci USA 92:3864–3868

Rayment I, Smith C, Yount R (1996) The active site of myosin. Annu Rev Physiol 58:671–702

Roopnarine O, Leinwand LA (1998) Functional analysis of myosin mutations that cause familial hypertrophic cardiomyopathy. Biophys J 75:3023–3030

Sata M, Ikebe M (1996) Functional analysis of the mutations in the human cardiac beta-myosin that are responsible for familial hypertrophic cardiomyopathy. Implication for the clinical outcome. J Clin Invest 98:2866–2873

Schroder RR, Manstein DJ, JahnW, Holden H, Rayment I, Holmes KC, Spudich JA (1993) Three-dimensional atomic model of F-actin decorated with *Dictyostelium* myosin S1. Nature 364: 171–174

Smith CA, Rayment I (1996) X-ray structure of the Magnesium(II)-pyrophosphate complex of the truncated head of *Dictyostelium discoidium* myosin to 2.7-Å resolution. Biophys J 70:1590–1602

Solomon SD, Geisterfer-Lowrance AAT, Vosberg HP, Hiller G, Jarcho JA, Morton CC, McBride WO, Mitchell AL, Bale AE, McKenna WJ, Seidman JG, Seidman CE (1990) A locus for familial hypertrophic cardiomyopathy is closely linked to the cardiac myosin heavy chain genes, CRI-L436, and CRI-L329 on chromosome 14 at q11-q12. Am J Hum Genet 47:389–394

Sweeney HL, Straceski AJ, Leinwand LA, Tikunov BA, Faust L (1994) Heterologous expression of a cardiomyopathic myosin that is defective in its actin interaction. J Biol Chem 269: 1603–1605

Thedinga E, McKenna WJ, Brenner B (1996) Search for functional defects resulting from the 606Val-Met mutation in the β-MHC gene associated with HCM. Biophys J 70:A41

Thierfelder L, Watkins H, MacRae C, Lamas R, McKenna WJ, Vosberg HP, Seidman JG, Seidman CE (1994) Alpha-tropomyosin and cardiac troponin T mutations cause familial hypertrophic cardiomyopathy: a disease of the sarcomere. Cell 77:701–712

Timson DJ, Trayer HR, Trayer IP (1998) The N-terminus of A1-type myosin essential light chains binds and modulates myosin motor function. Eur J Biochem 255:654–662

Vale RH (1996) Switches, latches, and amplifiers: common themes of G proteins and molecular markers. J Cell Biol 135:291–302

Vikstrom KL, Leinwand LA (1996) Contractile protein mutations and heart disease. Curr Opin Cell Biol 8:97–105

Wadgaonkar R, Shafiq S, Rajmanickam C, Siddiqui MAQ (1993) Interaction of a conserved peptide domain in recombinant human ventricular myosin light chain-2 with myosin heavy chains. Cell Mol Biol Res 39:13–26

Watkins H, Rosenzweig A, Hwang DS, Levi T, McKenna WJ, Seidman CE, Seidman JG (1992) Characteristics and prognostic implications of myosin missense mutations in familial hypertrophic cardiomyopathy. N Engl J Med 326:1108–1114

Watkins H, Thierfelder L, Anan A, Jarcho JA, Matsumori A, McKenna WJ, Seidman JG, Seidman CE (1993) Independent origin of identical beta cardiac myosin heavy-chain mutations in hypertrophic cardiomyopathy. Am J Hum Genet 53:1180–1185

Watkins H, Conner D, Thierfelder T, Jarcho JA, MacRae C, McKenna WJ, Maron BJ, Seidman JG, Seidman CE (1995a) Mutations in the cardiac myosin binding protein-C gene on chromosome 11 cause familial hypertrophic cardiomyopathy. Nat Genet 11:434–437

Watkins H, McKenna WJ, Thierfelder L, Suk HJ, Anan R, O'Donoghue A, Spirito P, Matsumori A, Moravec CS, Seidman JG, Seidman CE (1995b) Mutations in the genes for cardiac troponin T and alpha-tropomyosin in hypertrophic cardiomyopathy. N Engl J Med 332:1058–1064

Watkins H, Seidman JG, Seidman CE (1995c) Familial hypertrophic cardiomyopathy: a genetic model of cardiac hypertrophy. Hum Mol Genet 4:1721–1727

Coupling Between Chemical and Mechanical Events and Conformation of Single Protein Molecules

Yoshiharu Ishii[1], Akihiko Ishijima[2], and Toshio Yanagid[3]

1 Introduction

Actin and myosin interact and undergo a sliding movement, achieving muscle contraction and other cell motility in a living organism. For this mechanical activity the energy is derived from ATP hydrolysis. Biochemical studies have shown that there are a series of chemical reactions or states that make up the overall ATP hydrolysis scheme (Lymn and Taylor 1971). The basic chemical process includes ATP binding to myosin, hydrolysis of ATP to ADP and Pi, and the dissociation of Pi and ADP from myosin. Depending on the chemical state, the interaction of myosin with actin changes (Eisenberg et al. 1980). Binding of ATP to myosin weakens the interaction with actin and the release of Pi and ADP strengthen the interaction. Both X-ray and optical diffraction studies have suggested structural changes of myosin and actin, depending on the state of the actomyosin ATPase (Huxley 1969). Recently, crystallography combined with mutagenesis has suggested structural changes in the myosin molecule (Rayment et al. 1993; Dominguez et al. 1998). These studies have tempted investigators to link one chemical state with the unique structure of myosin, which in turn determines the interaction with actin and the mechanical events in a sequential manner. The hypothesis of 1:1 coupling among these events seems simple. However, this coupling must be tested experimentally.

In fact, there are many experimental results that suggest that myosin molecules move a long distance with the hydrolysis of a single ATP molecule (Harada et al. 1990; Higuchi and Goldman 1991; Ishijima et al. 1991, 1994, 1996; Lombardi et al. 1992). If chemical, structural, and mechanical events are tightly coupled, the myosin molecule can only move by a single step per ATP hydrolyzed, which will be no bigger than the size of the myosin molecule itself (8–10 nm). Data suggesting that myosin can step a long distance per ATP have driven us to rethink the mechanism of muscle contraction. The most unambiguous and direct answer to this question would be obtained by directly determining the coupling between the chemical and mechanical events. These experiments must be at the single molecule level, otherwise signals are

[1] Single Molecule Processes Project, ICORP, JST, 2-4-14 Senba-Higashi, Mino, Osaka 562 Japan
[2] Applied Physics, Dept. of Engineering, Nagoya University, Furo-cho, Chikusaku, Nagoya 464 Japan
[3] Dept. of Physiology, Osaka University Medical School, 2-2 Yamadaoka, Suita, Osaka 565 Japan

Results and Problems in Cell Differentiation, Vol. 36
D.D. Thomas, C.G. dos Remedios (Eds.): Molecular Interactions of Actin

averaged over many molecules. Results averaged over many molecules are complicated and difficult to interpret.

2 Measurements of Force Exerted by Single Myosin Molecules

The force produced by single myosin molecules can be measured by either laser trap or glass microneedle methods (Kishino and Yanagida 1988; Ishijima et al 1991; Svoboda et al. 1993; Finer et al. 1994; Molloy et al. 1995;). The two measurements share the premise that the force is determined by measuring the displacement utilizing spring-like properties of the laser trap or microneedle. For the simultaneous measurements, we used the laser trap (Ishijima et al. 1998; Tanaka et al. 1998). A bead is trapped by laser when it is placed on its focal point. If the trapped bead is pulled out of the focal point by a load from outside, the bead moves until the load is matched by the restoring force exerted by the trap. The magnitude of the force is proportional to the displacement of the bead and the spring constant. Thereby, we can determine the force by measuring the displacement. To measure the force exerted by myosin on an actin filament, the actin filament is suspended between two beads held by two independent laser traps. The actin filament is then positioned to interact with myosin molecules (see Fig. 1). The interaction with single myosin molecules in the presence of ATP produces the displacement of the bead, and bead displacement can be measured with nanometer accuracy using a quadrant photodiode detector.

We have used single myosin molecules in myosin-myosin rod co-filaments or subfragments attached to an artificial surface to avoid possible problems including: (1) effects of random orientation of myosin heads relative to an actin filament, and (2) effects caused by the direct attachment of myosin to the artificial surface. We have demonstrated that the interaction of myosin with actin is strongly dependent on the angle between the two filaments (Tanaka et al. 1998). The bead displacement is maximum when two filaments are aligned in parallel as in muscle. It decreases with an increase in the angle between two filaments, approaching zero when two filaments are perpendicular one another (Fig. 2). The average displacement would be in between a maximum and a minimum, if the measurements were done with myosin molecules that are randomly oriented. Filaments have the advantage of a known orientation of the myosin head. Another possible artifact is an effect of direct attachment of the myosin subfragment to the artificial surface. When S1 is directly attached to the glass surface, the velocity of actin filaments produced by S1 is slower than that produced by whole myosin. The whole myosin is most likely attached to the glass surface through its rod portion. When the myosin subfragment is attached to the artificial surface via biotin-avidin and light chain, which minimize the effects of the artificial surface, the velocity of actin filaments is restored to the same level as for myosin (Iwane et al. 1997). Thus, direct attach-

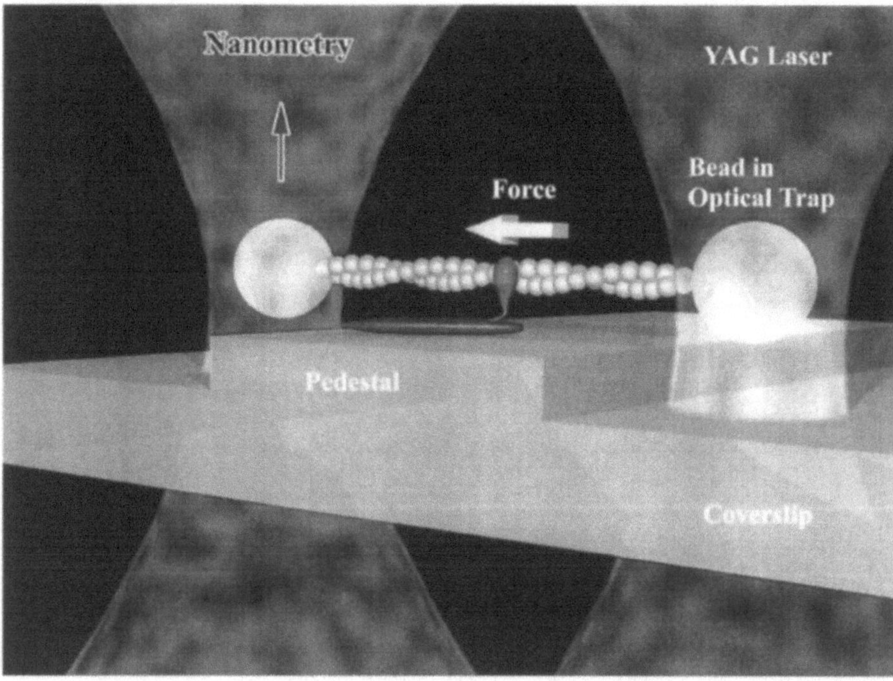

Fig. 1. Schematic drawing of the system used for the laser trap measurement. An actin filament is suspended between two beads held by a laser trap at each end. The displacement generated by the interaction with a single myosin head in the presence of ATP was determined by measuring the displacement of one bead with nanometer accuracy. The force exerted is calculated from the displacement, since it is proportional to it

ment of proteins to the artificial surface may affect their function and this problem is largely eliminated by use of myosin molecules in co-filaments with myosin rods.

To make sure that only a single myosin head interacts with the actin filament at any one time, the number of myosin heads along a myosin filament must be reduced. We have synthesized myosin filaments by mixing myosin with a large excess of myosin rods (Ishijima et al. 1998; Tanaka et al. 1998). Two heads of myosin may make the interpretation more complicated. Thus, one-headed myosin has been prepared instead of two-headed myosin. Mixing of one-headed myosin with myosin rods at a molar ratio of 1:1000 gave a filament of 5–8 μm containing only one or two heads.

The bead displacement is measured with an accuracy of a nanometer. The displacement measured is not of the myosin head itself, but of the bead attached to actin filaments. Actual displacement of the myosin head may be greater than that of the bead, because of any compliance between myosin molecules and the bead. Displacement must be corrected for any compliance within the myosin head, actomyosin interface, actin and actin/bead attachment.

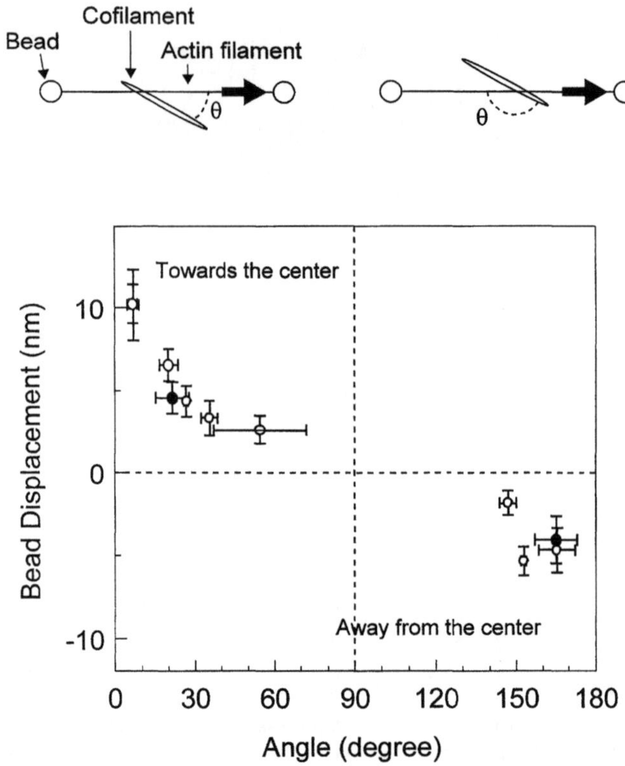

Fig. 2. Orientation dependence of the displacements. The displacements were measured at various angles between an actin filament and a myosin-rod co-filament. At 0° (the filaments are oriented in parallel), the displacements were maximum and then decreased with an increase in the angle

Our results show that the displacement produced by single myosin molecule is 15–20 nm, greater than the size of myosin molecule (Fig. 3). As shown by the simultaneous measurements described below, this displacement is produced by the breakdown of a single ATP molecule. This means that several mechanical events must take place during hydrolysis of a single ATP molecule. In fact, recently, we have succeeded in monitoring the details of the force development process with the probe microscope (Kitamura et al. 1999). A single force production process induced by the breakdown of a single ATP molecule is made up of many sub-steps of 5.5 nm, which reflects an interval between the nearest neighbor actin monomers.

3 Visualization of Turnover of Single ATP Molecule

The hydrolysis of a single ATP molecule can be monitored using fluorescent analogues of ATP and visualizing its binding to single myosin molecules

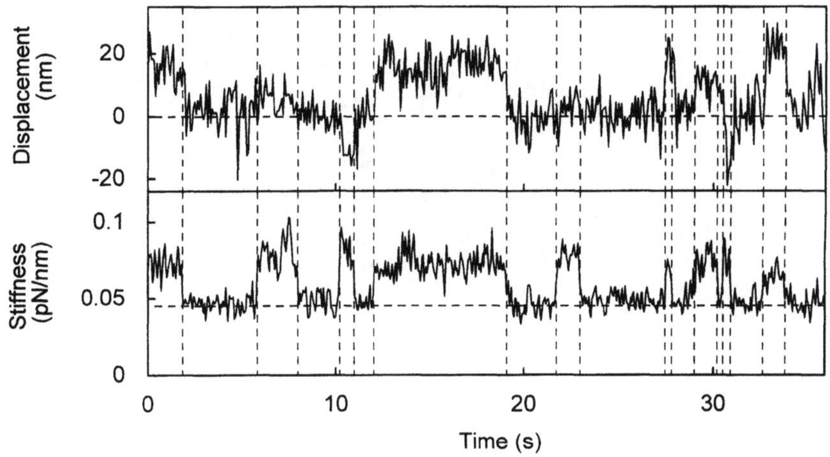

Fig. 3. The displacements generated by one-headed myosin. Typical records of the displacements with time are shown. The displacements were greater than the size of the myosin head. The *bottom panel* is the stiffness, a measure of the interaction of myosin and actin, which was determined from the variance of the Brownian motion of the bead. Since ATP concentration was low (100 nM), the duration of the mechanical events were long. Otherwise, the force and the thermal noise could not be distinguished

(Funatsu et al. 1995). Imaging of single fluorescent dye molecules in an aqueous solution was required for these experiments.

Seeing is believing. More than a decade ago a single actin filament with fluorescently labeled phalloidin was visualized and the sliding movement produced by myosin molecules adsorbed on the slide glass was monitored under the microscope (Yanagida et al. 1984; Kron and Spudich 1986; Harada et al. 1987). We recognized, for the first time, that the sliding motion occurs even at the molecular level. The next challenge for the technology was visualization of single motor molecules. At the beginning of the 1990s, single fluorescence dye molecules were visualized on an air-dried surface (Betzig and Chichester 1993), but not in aqueous solution where the fluorescence is quenched by water molecules to a large extent. Since video cameras already had the capacity to detect a single fluorescent dye molecule, even in aqueous solution, most of the difficulty was reducing the background noise. To reduce this background noise, we use local illumination; we irradiate only what we want to see. This has been achieved using an evanescent field produced by total internal reflection of laser light. An evanescent field is generated only with a depth of 100 nm from the interface between the slide glass and the sample solution (Fig. 4). Only protein molecules that are located on the surface of the slide glass and within the small area that the evanescent light illuminates can be visualized. In addition to the development of this microscope, we have searched for good fluorescent dyes for single molecule detection. Cyanine dyes are the best for single molecule detection, because they are bright and photo-chemically stable.

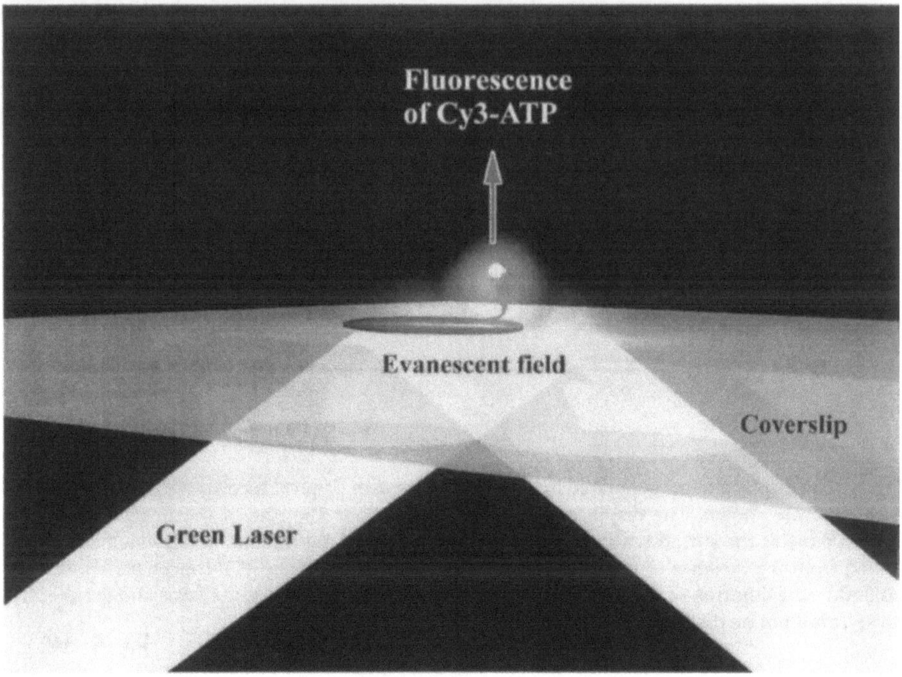

Fig. 4. Schematic drawing of evanescent illumination. Only fluorescence dyes that are located within the vicinity of the surface of the solution/coverslip are excited by the evanescent light that is generated by total internal reflection of the green laser. This local illumination reduces the background level to a large extent, facilitating the visualization of single dye molecules. This picture shows how to obtain an image of a fluorescent analogue of ATP bound to a single myosin head

With this technique, we succeeded in visualizing single fluorophores attached to protein molecules.

The evidence that a single spot arises from a single dye molecule is obtained from observed photobleaching. After irradiation by laser, the fluorescence intensity drops to the baseline in a single step. This is in contrast to bulk measurements in which many molecules are included and photobleaching occurs gradually. Additional evidence has been obtained from a comparison of fluorescence images with electron microscope images (Funatsu et al. 1995). Further, it has been demonstrated that the single spots of fluorescently labeled kinesin molecules move along microtubules, indicating that fluorescently labeled molecules still maintain their function (Vale et al. 1996).

To visualize the hydrolysis of single ATP molecules with this single molecule imaging technique, ATP is chemically labeled with fluorescent dye, Cy3, at its ribose (Funatsu et al. 1995; Tokunaga et al. 1997). The fluorescent dye attached to the ATP does not interfere with either binding to myosin or its ability to be hydrolyzed by myosin. The fluorescent ATP can be detected as a spot when it is immobile, as when it attaches to myosin molecules that are fixed to the glass surface. However, we cannot detect a spot of the fluorescent ATP when it undergoes Brownian motion, i.e., when it is not associated with myosin. For the experiments, myosin is labeled with Cy5 and sparsely attached onto the slide glass. The positions of the myosin molecules are marked and then the laser is switched to green to excite Cy3-ATP; the fluorescence increases when ATP binds to myosin and disappears when it dissociates from myosin. Thus, we can monitor the association of ATP to myosin and its dissociation, most likely after it is hydrolyzed to ADP. To confirm that this is due to ATP turnover, the time interval that the nucleotide stays bound to myosin is measured for individual ATP molecules and compared with the ATPase rate of myosin S1 in solution. Since this process is stochastic, distribution of the time interval can be fitted to an exponential curve. The decay rate constant was consistent with the ATPase rate of myosin S1. The possibility that ATP attaches and detaches without hydrolysis, and that fluorescent ATP molecules are photobleached before they are dissociated from myosin, is negligible.

4 Coupling Between Chemical and Mechanical Events

Since displacement and ATPase can now be measured at the single molecule level, we have combined these two techniques to measure both events simultaneously (Ishijima et al. 1998).

In Fig. 5, the configuration of myosin and actin filaments is depicted in the top panel and the fluorescence image of the two filaments labeled with Cy5 is shown in the second panel. The images of Cy3-nucleotides, sequentially taken, are shown in the lower panels. Figure 6 shows traces of the mechanical and biochemical events recorded simultaneously. The top panel shows the displacement and the bottom panel the fluorescence of Cy3-ATP at the position of a single myosin head. It is worth mentioning repeatedly that the increase and decrease in fluorescence intensity is due to the association and dissociation, respectively, of a single ATP molecule to the myosin head. The middle is the stiffness change, a good measure of the interaction of myosin with actin, which is determined from the variance of displacement due to Brownian motion of the bead. When myosin interacts with actin, the Brownian motion is suppressed and the stiffness increases. At low concentration of Cy3-ATP (10–20 nM), displacement driven by the hydrolysis of Cy3-ATP is maintained for tens of seconds until a new Cy3-ATP molecule binds to the myosin head. The decrease in the displacement and the stiffness occurs due to the detachment of the myosin head from the actin filament upon the binding of a new Cy3-ATP. Changes in displacement and stiffness are associated with increase in the fluorescence intensity in this case. Changes in the displacement and an increase in the stiffness are also associated with a decrease in the fluorescence intensity, or association of a Cy3-nucleotide. Thus, the mechanical events,

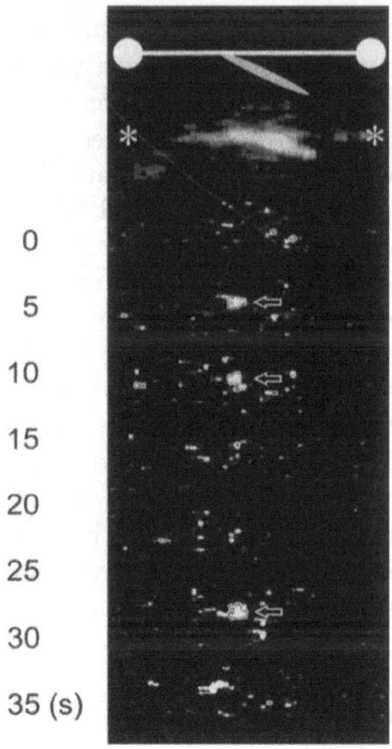

Fig. 5. Images of ATPase turnover by single myosin heads. When the fluorescent ATP is bound to a single myosin head, the fluorescent spot can be observed (indicated by *arrow*) and disappears when the ATP molecule dissociates from the myosin head. At the *top* is a schematic picture of an actin filament attached to beads at both ends, and a myosin-rod co-filament. The rest of the figure shows their fluorescence images

roughly speaking, appear to be associated with ATPase reactions at the time resolution of 1/30 s.

Since the measurement system worked well, we improved its time resolution to determine the coupling of chemical and mechanical events with a higher resolution. Figure 7a shows the process triggered by the binding of ATP on an expanded scale (1/120 s). When a Cy3-ATP nucleotide binds, the myosin head rapidly detaches from actin. The spread of timings between the Cy3-ATP binding and the detachment of the myosin head is only within 0.05 s (Fig. 7b). Thus, the change in the interaction between myosin and actin is coupled with the binding of ATP.

Myosin heads reattach to the actin filament and generate force. The force generations are triggered by the release of the bound Cy3-nucleotide, most likely Cy3-ADP. Figure 8a, b shows these processes on an expanded time scale (1/120 s). In approximately 50% of cases, the force generation occurs at the same time as the release of Cy3-ADP, within an experimental accuracy (Fig. 8a). In other cases, however, the force generation is delayed as much as 1 s after dissociation of Cy3-ADP (Fig. 8b). Figure 8c shows the distribution of timing of the force generation after the release of Cy3-ADP from the myosin head. Thus, the mechanical events are not always coupled with the chemical reaction.

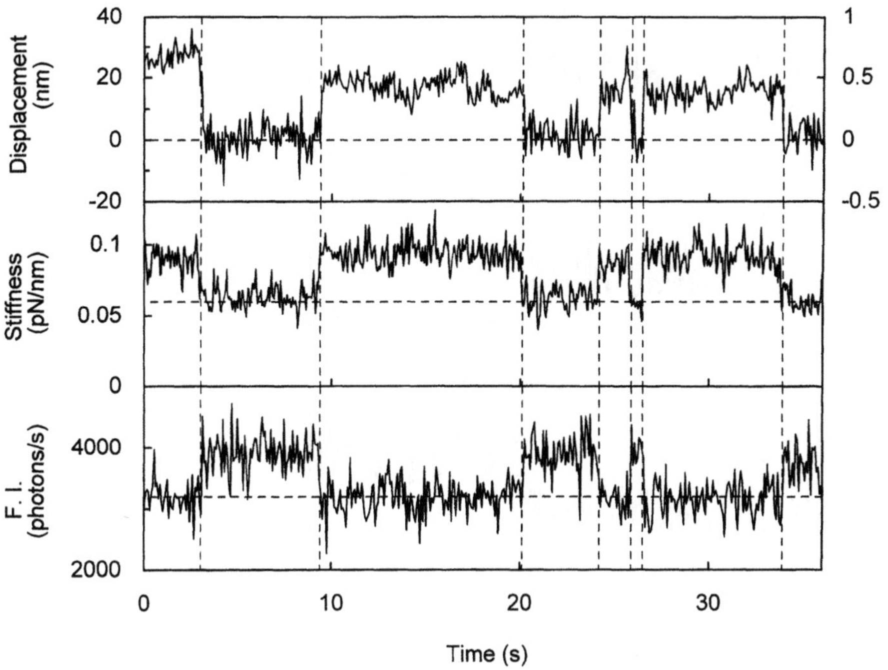

Fig. 6. Typical data from the simultaneous measurement of ATPase turnover and force generation by a single myosin head. The *top panel* shows the displacement of the bead attached to the actin filament, the *middle panel* shows stiffness as a measure of the actomyosin interaction, and the *bottom panel* shows fluorescence changes of fluorescently labeled ATP, a measure of the ATPase turnover. The mechanical events seemed coupled to the chemical events at this time range

The original "swinging cross-bridge model" or renovated "lever arm model" were proposed based on the results of both mechanical and structural studies. The elementary steps of the actomyosin-ATPase reaction have been characterized using isolated proteins in solution and the kinetic pathway has been related to the swinging cross-bridge model. Recently, the three-dimensional atomic structure of a myosin head has been revealed, which allows a more detailed analysis of the model in terms of the structure. In the absence of actin, the rate-limiting process is a release of Pi from myosin and this process is greatly accelerated by the presence of actin, resulting in an actin-induced activation of the myosin ATPase. In addition, the free energy greatly decreases upon release of ADP and Pi from actomyosin. These results suggest that force develops on release of Pi and/or ADP. However, our experiments show that the force generation and association of myosin to actin are delayed after the release of ADP. The energy must be stored in the myosin and used for generation of the force afterwards. It has been known that some enzymes, termed hysteretic

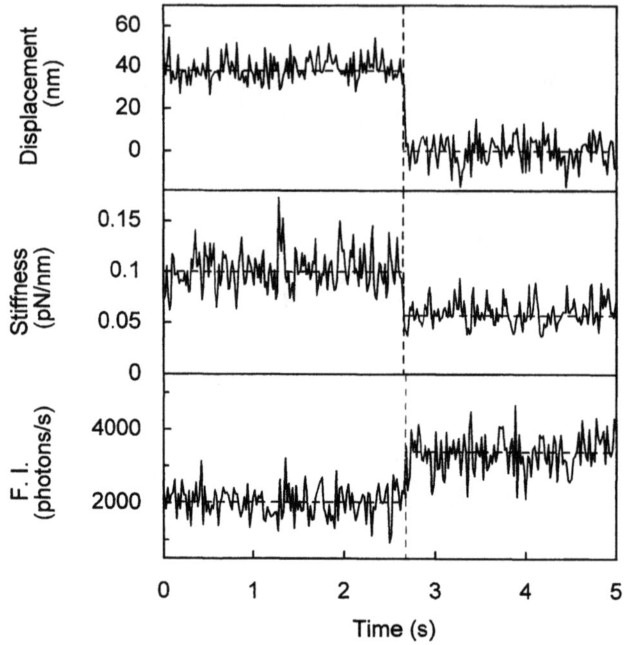

a

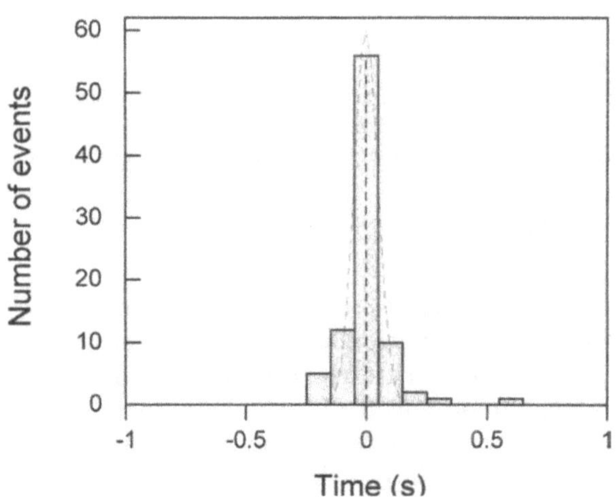

b

Fig. 7a,b. ATP binding and the dissociation of single myosin molecules. **a** The time course of the dissociation of single myosin molecules from the actin filament induced by the binding of ATP is shown on an expanded scale. Two events occurred at the same time within experimental accuracy. **b** The total 87 events are summarized in this histogram

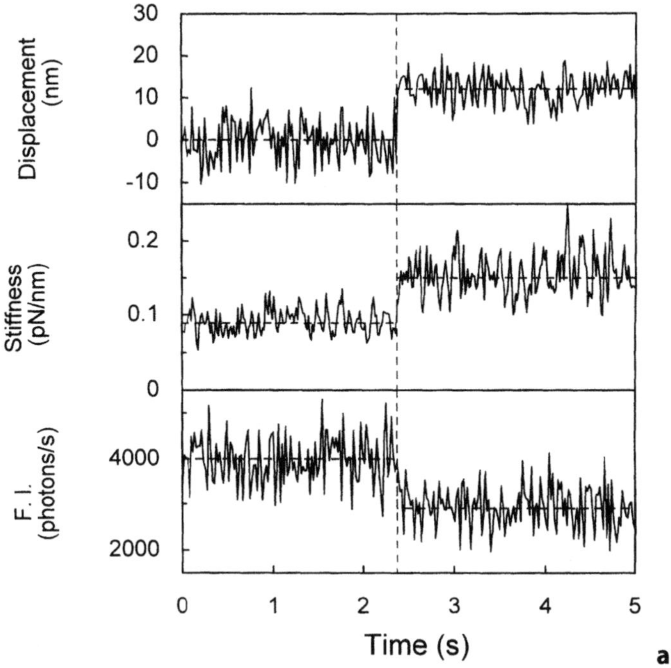

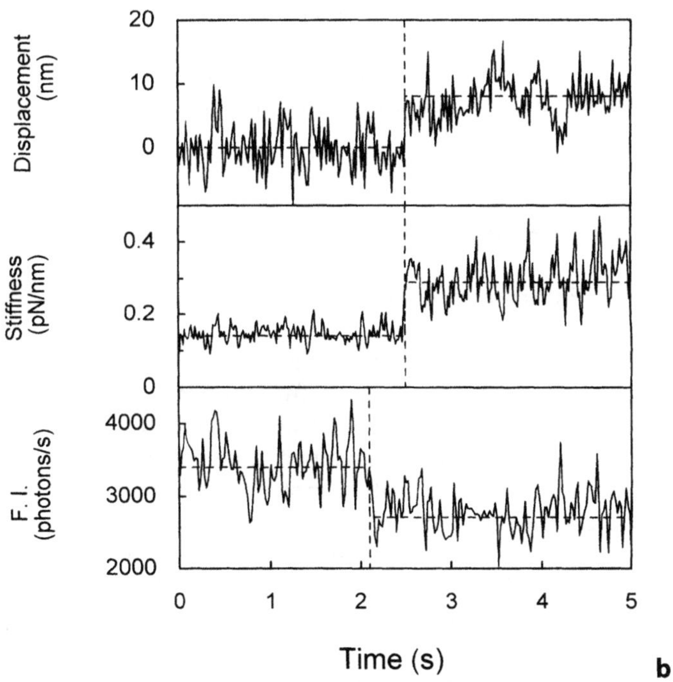

Fig. 8a–c. Nucleotide dissociation and force generation. The time course of nucleotide dissociation and force generation events is shown in an expanded scale. In about half of the cases, the chemical and mechanical events occur at the same time (**a**). However, in the rest of the cases, the force generation is delayed after dissociation of the nucleotide (**b**). The total 85 events are summarized in the histogram (**c**) (**c** see page 98)

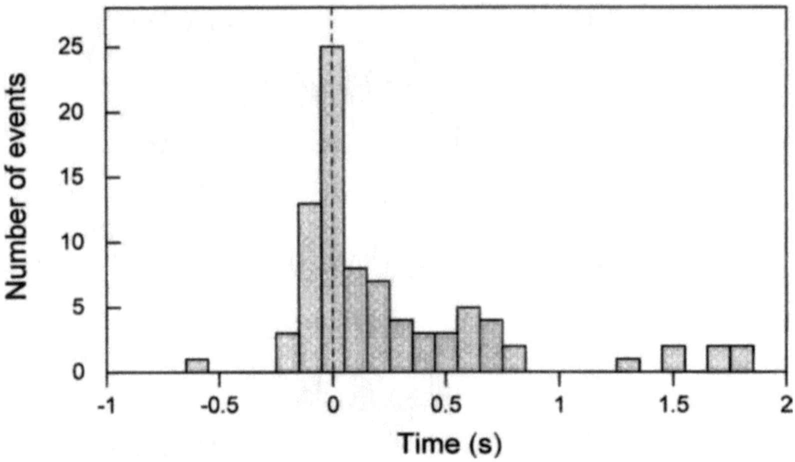

Fig. 8a–c. *Continued*

enzymes, respond slowly over seconds or minutes to rapid changes in ligand concentration. Kinetic analysis has suggested that the enzymes can exist in two or three different conformational states and slow transitions among these states lead to the hysteretic behavior (Frieden 1970). How could this be possible? To check for this possibility, we need to monitor the dynamic behavior of motor proteins.

5 Detection of Conformation of Single Protein Molecules

5.1 Fluorescence Resonance Energy Transfer of a Single Protein Molecule

Protein molecules have complicated ternary structures and their dynamic behavior may be crucial for function. We would like to determine the changes that take place within the protein molecules. To this end, we have focused our efforts on the development of a technology that can detect conformational changes of single protein molecules. In bulk experiments it is impossible to detect the dynamic behavior of individual single protein molecules because of signal averaging over many protein molecules. To observe the conformational states of single protein molecules, the fluorescence spectroscopic technique and the single molecule imaging technique have been combined (Ishii et al. 1997).

Among the spectroscopic techniques, fluorescence resonance energy transfer (FRET) is a ruler that can measure the distance between two different fluorescent probes (Stryer 1978). When the donor and acceptor are located close to each other, the excited energy of the donor is transferred to the acceptor without radiation, and then the acceptor fluoresces. When the donor and acceptor are located apart, only the donor fluoresces. The efficiency of the FRET depends on the distance as $1/r^6$, where r is the distance between donor and acceptor. Thus, we can describe the specific geometry of proteins by monitoring the color of the fluorescence. To perform this experiment at the single molecule level, Cy3 or tetramethylrhodamine (TMR) is chosen as a donor and Cy5 as an acceptor. The fluorescence emission spectrum of Cy3 (or TMR) and the absorption spectrum of Cy5 overlap well and the fluorescence bands of Cy3 and Cy5 are well separated for good results at the single molecule level. The Förster distance, a distance that gives the FRET efficiency 50%, is approximately 5.5 nm.

SH1 and SH2 of myosin S1 are labeled with Cy3 and Cy5, respectively. The most primitive way to monitor FRET is dual-color imaging at the wavelengths of the donor and acceptor emission. When the donor is excited, the images of donor and acceptor are monitored simultaneously (Fig. 9). Using the fluorescence spectrum of single protein molecules, we confirmed that these images really reflect FRET (Ishii et al. 1997). To do this, only the fluorescence of a single spot is exclusively passed through a pinhole into a spectroscopic system. The fluorescence spectrum image is taken with a back-illuminated cooled CCD camera. A typical fluorescence spectrum of S1 dually labeled with Cy3 and Cy5 at SH1 and SH2, respectively, is shown in Fig. 10. As expected, there are two peaks observed at the Cy3 and Cy5 peak positions. Further evidence that this is a FRET spectrum from a single molecule is obtained from photobleaching experiments. One of the two, donor or acceptor, is photobleached first. The spectral changes upon photobleaching depend on which fluorophore photobleached first. When the donor is photobleached first, both donor and acceptor fluorescence disappear. When the acceptor is photobleached first, the donor fluorescence increases, with an associated drop in

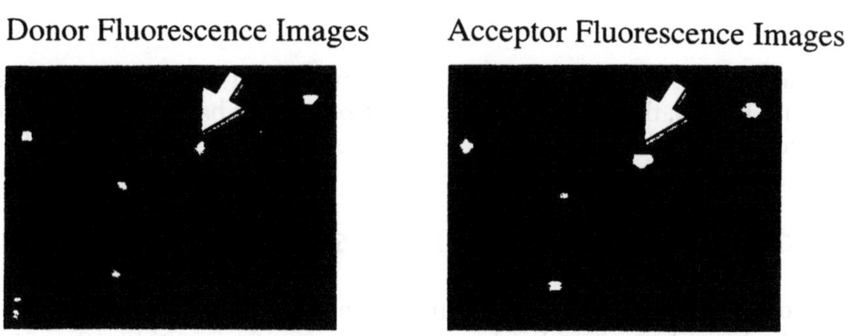

Fig. 9. Images of the donor and acceptor fluorescences of individual protein molecules. Protein molecules were labeled with tetramethylrhodamine as a donor and Cy5 as an acceptor. Labeled protein molecules were adsorbed onto the glass surface. Fluorescence images of the donor and acceptor were simultaneously taken when the donor was excited. The acceptor fluorescence images are due to the fluorescence resonance energy transfer (FRET) from the donor. Single spots arise from single protein molecules

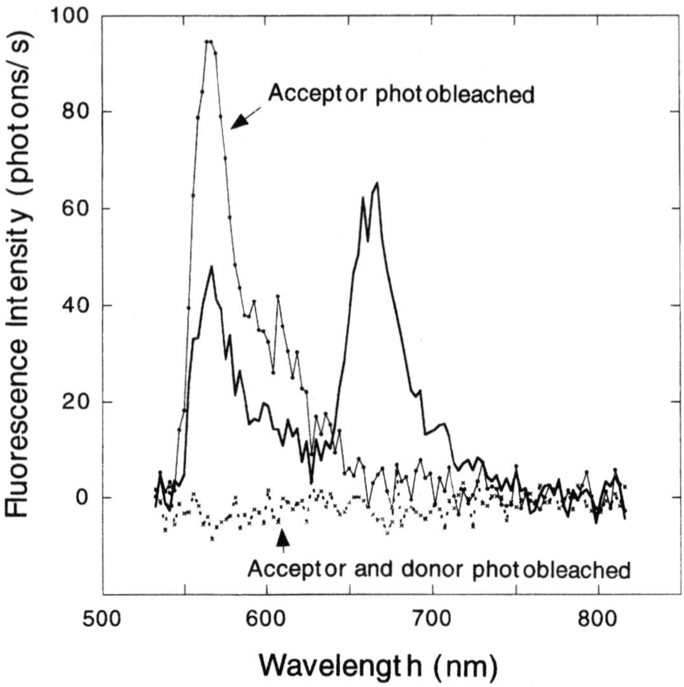

Fig. 10. The fluorescence spectra from single S1 molecules. The single molecule imaging microscope has been further developed to take fluorescence spectra from single molecules. The spectrum showed two peaks at the position of the tetramethylrhodamine and Cy5 band. The florescence spectrum was taken every half second and the photobleaching processes were monitored. In this case, the acceptor was first photobleached, and then the donor was photobleached. Upon photobleaching of the acceptor, the acceptor fluorescence dropped to the baseline and the donor fluorescence increased. Then the donor fluorescence dropped to the baseline due to the photobleaching of the donor

the acceptor fluorescence, since no quenching of the donor fluorescence occurs. Finally, the donor fluorescence drops to the baseline when the donor is photobleached.

5.2 Conformational Diversity of Myosin Subfragments

The donor and acceptor fluorescence intensity of dually labeled myosin S1 is largely dependent on molecules in the absence of nucleotides. One S1 molecule has high donor fluorescence and low acceptor fluorescence, whereas another S1 molecule has low donor fluorescence and high acceptor fluorescence. This suggests that the relative geometry between the donor and acceptor changes from molecule to molecule. As a summary, the data of individual protein molecules are plotted as the donor intensity vs acceptor fluorescence

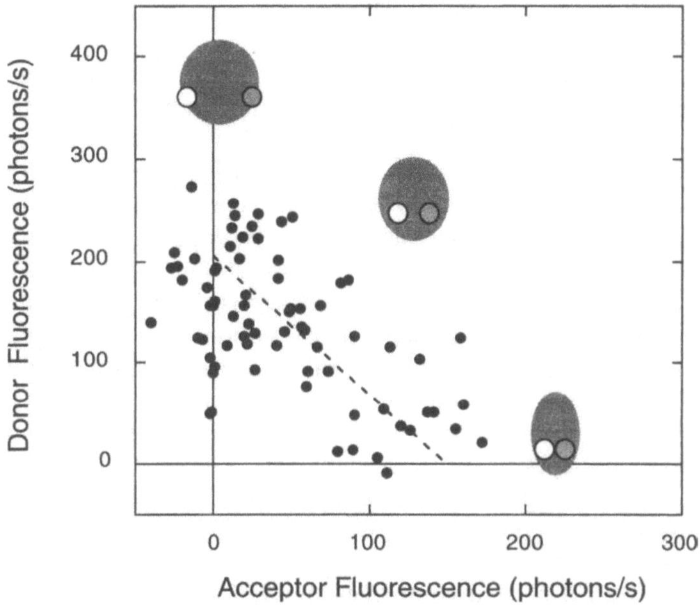

Fig. 11. The diversity of the FRET of individual S1 molecules. The fluorescence intensities of the donor and acceptor of individual S1 molecules are measured from the images in Fig. 9. The data are summarized by plotting the donor fluorescence intensity for one axis and the acceptor fluorescence intensity for the other axis. The data points were distributed over a wide range, indicating that individual molecules have different FRET efficiencies or conformations

intensity (Fig. 11). Each point represents data from a single molecule. From our current understanding of protein structure, we would expect a constant FRET efficiency and the data should all overlap at a single point. However, the experimental points are distributed on a line that crosses both axes, confirming that the donor and acceptor fluorescence depends on molecules in such a way that one increases proportional to the amount of the other. The point at which the best-fitted line crosses the axis of the donor intensity must be a point that corresponds to the FRET efficiency at 0%, and the point at which it crosses with the axis for the acceptor intensity must be a point that corresponds to the FRET efficiency at 100%. Thus, the distance between SH1 and SH2 varies from molecule to molecule.

The question has arisen as to how such a variety of protein conformations of different molecules could be possible. One possible explanation is that the protein conformation varies slowly. To check if this was the case, the time dependence of the fluorescence was measured. The fluorescence of the donor and acceptor spontaneously fluctuate with time in a correlated manner. Basically, the donor fluorescence increases with a decrease in the acceptor fluorescence and decreases with an increase in the acceptor fluorescence, indicating that the FRET efficiency fluctuates over a time range of seconds. The slow fluctuation of the FRET efficiency explains the diversity among molecules at any given time.

Slow conformational transition has also been indicated by the spectral change of the single fluorophore attached to myosin S1 under the same experimental conditions (Wazawa et al. 2000). As described above, the fluorescence spectrum of TMR attached to single protein molecules could be measured. A series of spectra are taken every half second and the peak position of the spectrum is determined for each one. The peak position of single TMR fluorophore attached specifically to SH1 of myosin S1 fluctuated within 2–3 nm on the time scale of seconds, whereas the spectral position of the dye molecules themselves varied little. The slow transition of the conformation of proteins is most likely responsible for changes in the microenvironment of the fluorophore attached to the protein. Measurements of the fluorescence spectrum in different solvents show that the spectrum of a single molecule is sensitive to solvents, as in bulk measurements.

The slow, spontaneous conformational transitions may explain the mechanism for the storage and release of energy in the actomyosin complex and the hysteretic enzymes. The energy released from the hydrolysis of ATP is stored in a certain conformation even after ADP dissociates, and then the energy is released with a spontaneous conformational transition. As will be mentioned below, actin molecules in the filamentous form also show spontaneous fluctuation on the time scale of seconds. Thus, the energy liberated from the ATP hydrolysis by myosin is stored in myosin or actin. Not only does the energy of ATP hydrolysis power more than a single myosin step, it can be stored within the S1 head to be released after myosin steps along actin. The coupling of the structural changes with the mechanical and chemical events must be studied.

5.3 Dynamic Behavior of Protein Molecules

This slow transition of FRET efficiency has also been obtained in other proteins such as between different subunits of troponin (Cys-133 of troponin I and Cys-98 of troponin C; Ishii et al. 1997) and different domains of actin (Gln-41 and Cys-374; Yokota et al. 1998). In these proteins, but with different donor and acceptor pairs, the distribution of the FRET efficiency has been indicated by the bulk lifetime measurements. In bulk measurements, averaging over both molecules and time are involved. Slow conformational transition, as observed in single molecule measurements, is in part responsible for the distribution of the FRET efficiency obtained in bulk measurements.

From measurements of these proteins, the general features of the fluctuation in the FRET efficiency are: (1) that these transitions occur mainly between two conformational states, (2) that the fluorescence change due to the transition occurs in a single step with experimental accuracy of milliseconds, and (3) that the transition occurs in the time scale of subseconds to seconds.

Since many proteins have shown slow changes in FRET efficiency, one may ask if the change is due to artifacts of the measurements rather than conformational changes of proteins. In the case of tropomyosin, the FRET efficiency between Cy3 and Cy5 attached to corresponding cysteine residues of two identical subunits of a coiled-coil stays constant. Thus, the conformation of tropomyosin does not change on the time scale in which some proteins show fluctuation.

One may ask how much structural change occurs associated with these slow conformational transitions. As we described above, the change in the distance can be calculated from the change in the FRET efficiency. However, we must note that not only the distance, but also the relative orientations of the two fluorophores determine the FRET efficiency. In bulk measurements, we usually assume that the fluorophore rotates very quickly and is fully randomized at the time range of the lifetime of the fluorophores. It is not easy to study how orientation of the donor and acceptor contributes to the FRET efficiency. All we could measure is each anisotropy of donor and acceptor at the single molecule level. In the case of S1, the anisotropy of the fluorophore attached to SH1 shows that the transition from one value to the other occurs at a time scale of seconds (Saito et al. 1997). This is consistent with the idea of slow conformational transition. This is an exceptional case. Usually the orientation of fluorophores of TMR, Cy3, and Cy5 that have been used for single molecule FRET imaging as described above is fully randomized, so the anisotropy stays at low values, without any fluctuation, and the FRET efficiency still shows fluctuations. When fluorophores are fully randomized, the fluctuation of the FRET efficiency can be attributed to the change in the distance between the donor and acceptor.

Multiple metastable conformations separated by barriers with high activation energy and the slow transitions among them have been suggested in many proteins. Multiple metastable conformational states of proteins have been indicated to occur during protein folding (Wang et al. 1996; Clark et al. 1997; Sohl et al. 1998). The GTP binding protein, ras, has been shown to exist in multiple conformations with slow transitions among them (Ito et al. 1997). These slow conformational transitions may be related to the multiple tasks for this protein during signal transduction. For an extreme case of slow transition, prions show very slow conformational transitions, which are thought to cause fatal diseases (Harrison et al. 1997).

6 Concluding Remarks

Many biomolecules use the energy liberated from the breakdown of ATP or other nucleotides. Molecular motor proteins generate force and displacement, whereas signal transduction proteins send messages to other molecules and channel proteins allow for the transport of material. For all systems, the coupling between the chemical reaction, structural changes and resultant function is important for understanding their mechanisms. Does every protein respond to the signal or input from the outside in a deterministic way as a lever arm model suggests? Or does every protein respond in variable ways such that myosin generates force with different timing after nucleotides are dissociated as suggested by the simultaneous experiments? Since proteins function with input energy similar to thermal energy, they must work in a manner different from man-made machines, which use energy much greater than thermal energy of the system. The answer to the question must be obtained experimentally. Experiments will show whether proteins are simply like man-made machines such as cogwheels, springs, and levers, that can be found in a toolbox. Or experiments will show whether proteins are something that we cannot replace with man-made machines. Searching for the answer is both intriguing and challenging.

The experiments described in this chapter were carried out by the members of Yanagida Biomotron Project and Single Molecule Processes Project and the students in Osaka University.

References

Betzig E, Chichester RJ (1993) Single molecules observed by near-field scanning optical microscope. Science 262:1422–1425

Clark AC, Raso SW, Sinclair JF, Ziegler MM, Chaffotte AF, Baldwin TO (1997) Kinetic mechanism of luciferase subunit folding and assembly. Biochemistry 36:1891–1899

Dominiquez R, Freyzon Y, Trybus KM, Cohen C (1998) Crystal structure of a vertebrate smooth muscle myosin motor domain and its complex with the essential light chain: visualization of the pre-power stroke state. Cell 94:559–571

Eisenberg E, Hill TL, ChenY (1980) Cross-bridge model of muscle contraction quantitative analysis. Biophys J 29:195–227

Finer JT, Simmons RM, Spudich JA (1994) Single myosin mechanics: piconewton forces and nanometre steps. Nature 368:113–119

Frieden C (1970) Kinetic aspects of regulation of metabolic processes: the hysteretic enzyme concept. J Biol Chem 245:5788–5799

Funatsu T, Harada Y, Tokunaga M, Saito K, Yanagida T (1995) Imaging of single fluorescent molecules and individual ATP turnovers by single myosin molecules in aqueous solution. Nature 374:555–559

Harada Y, Noguchi A, Kishino A, Yanagida T (1987) Sliding movement of single actin filaments on one-headed myosin filaments. Nature 326:805–808

HaradaY, Sakurada K, Aoki T, Yanagida T (1990) Mechanochemical coupling in actomyosin energy transduction studied by in vitro movement assay. J Mol Biol 216:49–68

Harrison PM, Bamborough P, Daggett V, Prusiner SB, Cohen FE (1997) The prion folding problem. Curr Opin Struct Biol 7:53–59

Higuchi H and Goldman YE (1991) Sliding distance between actin and myosin filaments per ATP molecule hydrolysed in skinned muscle fibres. Nature 352:352–354

Huxley HE (1969) The mechanism of muscular contraction. Science 164:1356–1366

IshiiY, Funatsu T, WazawaT, Yoshida T, Watai J, Ishii M, Yanagida T (1997) Communication between troponin-C and -I revealed by single molecule fluorescence spectroscopy and FRET. Biophys J 72:A283

Ishijima A, Doi T, Sakurada K, Yanagida T (1991) Sub-piconewton force fluctuation of actomyosin in vitro. Nature 352:301–306

Ishijima A, Harada Y, Kojima H, Funatsu T, Higuchi H, Yanagida T (1994) Single-molecule analysis of the actomyosin motor using nano-manipulation. Biochem Biophys Res Commun 199:1057–1063

Ishijima A, Kojima H, Higuchi H, Harada Y, Funatsu T, Yanagida T (1996) Multiple and single-molecule analysis of the actomyosin motor by nanometer-piconewton manipulation with a microneedle: unitary steps and forces. Biophys J 70:383–400

Ishijima A, Kojima H, Funatsu T, Tokunaga M, Higuchi H, Yanagida T (1998) Simultaneous measurement of chemical and mechanical reaction. Cell 92:161–171

Ito Y, Yamasaki K, Iwahara J, Terada T, Kamiya A, Shirouzu M, MutoY, Kawai G, Yokoyama S, Laue ED, Walchli M, Shibata T, Nishimura S, Miyazawa T (1997) Regional polysterism in the GTP-bound form of the human c-Ha-Ras protein. Biochemistry 36:9109–9119

Iwane AH, Kitamura K, Tokunaga M, Yanagida T (1997) Myosin subfragment-1 is fully equipped with factors essential for motor function. Biochem Biophys Res Commun 230:76–80

Kishino A, Yanagida T (1988) Force measurements by micromanipulation of a single actin filament by glass needles. Nature 334:74–76

Kitamura K, Tokunaga M, Iwane AH, Yanagida T (1999) Myosin step size determined by directly manipulating a single S1 molecule with a scanning probe. Nature 397:129–134

Kron SJ, Spudich JA (1986) Fluorescent actin filaments move on myosin fixed to a glass surface. Proc Natl Acad Sci USA 83:6272–6276

Lombardi V, Piazzesi G, Linari M (1992) Rapid regeneration of the actin-myosin power stroke in contracting muscle. Nature 355:638–641

Lymn RW, Taylor EW (1971) Mechanism of adenosine triphosphate hydrolysis by actomyosin. Biochemistry 10:4617–4624

Molloy JE, Burns JE, Kendrick-Jones J, Tregear RT, White DCS (1995) Movement and force produced by a single myosin head. Nature 378:209–212

Rayment I, Holden HM, Whittaker M, Yohn CB, Lorentz M, Holmes KC, Milligan RA(1993) Structure of the actin-myosin complex and its implications for muscle contraction. Science 261: 58–65

Saito K, Tokunaga M, Iwane AH, Yanagida T (1997) Polarization measurement of single fluorophores attached to motor protein in aqueous solution. Biophys J 72:A179

Sohl JL, Jaswal SS, Agard DA (1998) Unfolded conformations of α-lytic protease are more stable than its native state. Nature 395:817–819

Stryer L (1978) Fluorescence energy transfer as a spectroscopic ruler. Annu Rev Biochem 47: 819–846

Svoboda K, Schmidt CF, Schnapp BJ, Block SM (1993) Direct observation of kinesin stepping by optical trapping interferometry. Nature 365:721–727

Tanaka H, Ishijima A, Honda M, Saito K, Yanagida T (1998) Orientation dependence of displacements by a single one-headed myosin relative to the actin filament. Biophys J 75: 1886–1894

Tokunaga M, Kitamura K, Saito K, Iwane AH, Yanagida T (1997) Single molecule imaging and enzymatic reactions achieved by objective-type total internal reflection microscopy. Biochem Biophys Res Commun 235:47–53

Vale RD, Funatsu T, Pierce DW, Romberg L, Harada Y, Yanagida T (1996) Direct observation of single kinesin molecules moving along microtubules. Nature 380:451–453

Wang Z, Mottonnen J, Goldsmith EJ (1996) Kinetically controlled folding of the sperin plasminogen activator inhibitor1. Biochemistry 35:16443–16448

Wazawa T, Ishii Y, Funatsu T, Yanagida T (2000) Spectral fluctuation of a single fluorophore conjugated to a protein molecule. Biophys J 78:1561–1569

Yanagida T, Nakase M, Nishiyama K, Oosawa F (1984) Direct observation of motion of single F-actin filaments in the presence of myosin. Nature 307:58–60

Yokota H, Ishii Y, Wazawa T, Funatsu T, Yanagida T (1998) Dynamic structure of single actin molecules revealed by FRET imaging. Biophys J 74:A46

An Overview of Actin-Based Calcium Regulation

Sherwin S. Lehrer[1]

1 Introduction

Muscle contraction is initiated by an increase in intracellular Ca^{2+} concentration. In skinned muscle fiber preparations, where different solutions can be applied, the tension generated is sharply dependent on the pCa^{2+}. Yet it has become increasingly clear that the activation is not simply due to Ca^{2+} binding to the contractile apparatus, located on the thin filament, but that myosin heads, the tension generating motors originating from the thick filament, are involved in the cooperative response.

2 Cooperativity in the Ca^{2+} Regulation of Muscle Contraction

The possible sources of cooperativity in the regulation process are reviewed in the chapter by Geeves and Lehrer. The approach emphasizes the role of myosin heads in the activation process and the properties of Tm as effected by Tn in determining the cooperativity. A model calculation shows how Ca^{2+} can control the fraction of force generating cross-bridges.

What is emerging from structural, physiological and biochemical studies of regulation is the relationship of the weak complexes (A-states) and strong complexes (R-states) of myosin-actin with the three myosin binding states, blocked/closed/open (or two ATPase activity states, off/on) of the thin filament. In the blocked (off) state there is no productive myosin binding. In the closed (off) state there is binding to form the A-state but no force generation or activation of ATPase. In the open (on) state there is binding to form the R-state to generate force and activate ATPase.

It is the cooperative coupling of the R-state with the thin filament states, facilitated by Tn and Ca^{2+}, that determines the extent of activation. Thus, on binding of Ca^{2+} to troponin, sufficient weak binding complexes form to allow some strong complexes to be formed that cooperatively activate the thin filament, resulting in force generation and activation of ATPase.

Questions that need to be answered in order to gain a full understanding of regulation include: What is the high-resolution structure of each of the

[1] Muscle Research Group, Boston Biomedical Research Institute, Watertown, Massachusetts 02472, USA

Results and Problems in Cell Differentiation, Vol. 36
D.D. Thomas, C.G. dos Remedios (Eds.): Molecular Interactions of Actin

components in the different states?; What are the dynamic and equilibrium properties of the thin filament that determine the cooperative properties?; and What is the relationship between myosin binding and Ca^{2+} binding that determine activation?

3 Motility Assays of Calcium Regulation of Actin Filaments

In this chapter, La Madrid, Chase, and Gordon describe a recent application of the in vitro motility assay to understand how $[Ca^{2+}]$ regulates muscle contraction. They previously showed that in rabbit skeletal muscle, $[Ca^{2+}]$ grades the gliding speed of regulated thin filaments under physiological (140 mM) ionic strength conditions. At the $[Ca^{2+}]$ studied, it is found that although short filaments had a higher tendency to stop, when they were moving, their speeds were not significantly different from the long filaments. The possible implications are that calcium may be changing the strong binding time, thereby changing the duty ratio and/or the drag on the actin filament, in addition to controlling the number of motors that can bind.

4 The Ultrastructural Basis of Actin Filament Regulation

A low resolution model is available of the relationship between the components of the thin filament, actin, tropomyosin (Tm) and troponin (Tn), and of the structure of the myosin thick filament. Structural studies have suggested a model for the action of Ca^{2+} at the level of the thin filament. Thus, Ca^{2+} binding to troponin allows tropomyosin to move away from a site on actin, exposing a myosin binding site. This steric-blocking model has provided a framework for designing experiments. Recent reconstructions of electron micrograph images of reconstituted thin filaments have provided low-resolution information about movement of tropomyosin in regulation. The chapter by Craig and Lehman reviews these studies and provides evidence for a modified steric-blocking model that incorporates the influence of myosin heads.

5 The Role of Troponin in the Ca^{2+} Regulation of Skeletal Muscle Contraction

A great deal of information has been obtained from fibers and myofibrils by a variety of approaches, including selective extraction and exchange of components. The chapter by Szczesna and Potter, about fiber and solution studies, covers the roles of Tn components (TnC, TnI, and TnT) and suggests new functions for TnT in regulation.

6 Structural Changes Between Regulatory Proteins and Actin: A Regulation Model by Tropomyosin-Troponin Based on FRET Measurements

Solution studies have shown rearrangement of Tn components by Ca^{2+}, but little or no movement of Tm on the thin filament was detected using fluorescence resonance energy transfer (FRET) approaches. Although this could be due to lack of sensitivity of this approach for the Tm movement proposed in this system, a different mechanism of regulation involving actin is proposed in the chapter by Miki, based on FRET studies.

Cooperativity in the Ca^{2+} Regulation of Muscle Contraction

Michael A. Geeves[1] and Sherwin S. Lehrer[2]

1 Introduction

We have recently reviewed the evidence that the muscle thin filament operates as a classical Monod-Wyman-Changeux (MWC, 1965) cooperative/allosteric system while interacting with myosin heads (Lehrer and Geeves 1998). We showed that much of the available biochemical in vitro data can be explained if actin is considered to be the catalyst (enzyme) which accelerates the loss of Pi from the myosin-ADP-Pi substrate. In this scheme, tropomyosin (Tm) is the regulatory component, myosin complexed with ADP is the activating ligand and troponin (Tn) in the absence or presence of Ca^{2+} is the allosteric inhibitor or activator, respectively. The myosin product complex "turns-on" the activity by shifting the equilibrium from the T (closed) to the R-state (open), facilitated by Tn in the presence of Ca^{2+}. Thus, the ATPase activity of the system is mostly off in the absence of myosin heads, even in the presence of Ca^{2+}.

We introduced the concept of the apparent cooperative unit size, *n*, the number of actin subunits turned on by the strong binding of a myosin head and indicated that *n* can vary depending upon the strength of the end-to-end interactions and flexibility of Tm. We pointed out that in the absence of Ca^{2+}, there is significant population of another state, blocked, which is also involved in the regulation of the myosin-actin interaction. In the previous review we discussed the essential role of Tm in providing the cooperativity. In this review we will focus on the role of Ca^{2+} and extend our discussion to a consideration of the various interactions among the several components which are involved in the regulation of force generation in a muscle fiber.

2 The Problem: The Force/pCa^{2+} Curve

Muscle contraction is initiated by an increase in intracellular Ca^{2+} concentration. Although we know all of the essential components and the signal transduction pathway which leads from the change in free Ca^{2+} concentration to the contraction of a muscle fiber, the molecular details of the turning-on process

[1] Department of Biosciences, University of Kent, Canterbury, Kent, CT2 7NJ UK
[2] Muscle Research Group, Boston Biomedical Research Institute, Watertown, Massachusetts 02472, USA

Results and Problems in Cell Differentiation, Vol. 36
D.D. Thomas, C.G. dos Remedios (Eds.): Molecular Interactions of Actin

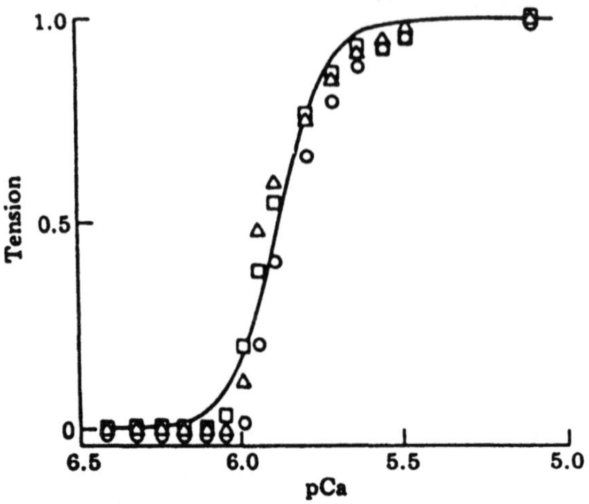

Fig. 1. Normalized isometric tension vs pCa. The line is a fit to a Hill plot with log K = 5.87 and Hill coefficient = 5.8. (After Brandt et al. 1980)

are poorly understood. One of the main reasons why this problem remains unsolved is the complexity of the interactions which involve six distinct protein subunits assembled into micrometer-long filaments that are themselves assembled into a precise three-dimensional array of interdigitating filaments. The potential for long-range cooperative interactions involving hundreds or thousands of molecules is clearly present. The complex cooperativity is illustrated in Fig. 1, which shows the Ca^{2+} dependence of the force developed by an isometrically contracting skinned muscle fiber.

The Hill coefficient for this Ca^{2+}-induced turning on of force is large (>5 for the example shown here but can range from 2.5 to >5 depending upon conditions) yet each troponin C binds only two Ca^{2+} ions. This large Hill coefficient suggests that cooperative binding of Ca^{2+} to TnC alone cannot account for the Ca^{2+} switch. In addition, the data in Fig. 1 as well as from other laboratories (Moss 1992), which show non-symmetrical force/pCa curves, appear to indicate a greater Hill coefficient at high pCa compared with low pCa. To understand this high cooperativity and non-symmetry we must first define the structure of the system.

3 Components of the Regulatory System

The muscle fiber in each half sarcomere consists of a set of interdigitating thick filaments and thin filaments in a hexagonal array. The relationship between the myosin of the thick filament and the actin of the thin filament in the overlap zone is illustrated in Fig. 2. The core of the thick filament is composed of the tails of myosin and other components including C-protein and titin. The myosin heads (cross-bridges or S1), project out from the backbone at regular

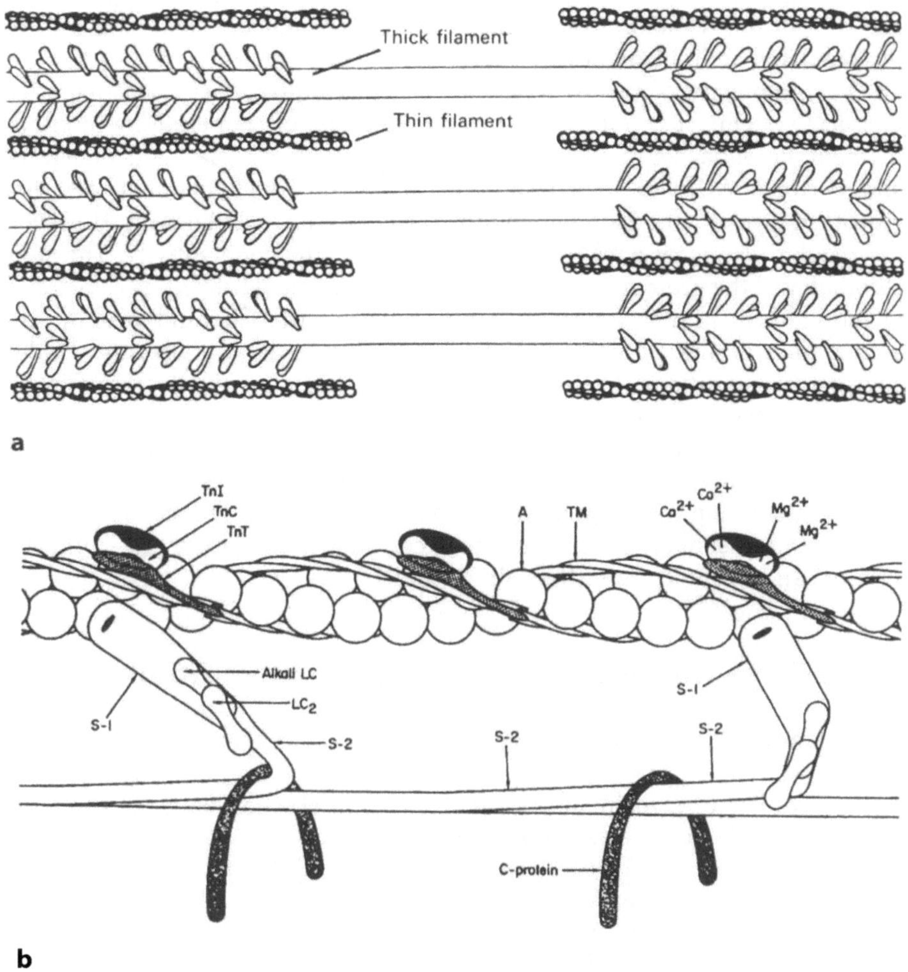

Fig. 2. a Schematic diagram of the components of the filament arrangement in the central region of the striated muscle. **b** Schematic diagram of the components of the thin and thick filaments of striated muscle. (After Hoffman et al. 1991)

intervals and can interact with actin in the thin filament. The thin filament consists of a polymer of globular actin subunits which form the characteristic arrangement of a twisted double string of pearls. Tropomyosin (Tm), a coiled-coil alpha helix, interacts with seven actin monomers and with itself head-to-tail. Each Tm is associated with one Tn consisting of three subunits; troponin I (TnI), troponin C (TnC) and troponin T, (TnT). TnC is the Ca^{2+} binding subunit, TnI is the inhibitory subunit which interacts with actin and TnC in a Ca^{2+}-dependent manner, and with TnT. TnT, an elongated molecule, interacts with both TnI and TnC through its C-terminal domain and with the head-to-tail region of Tm through its N-terminal domain. Other thin filament components include actin capping proteins and nebulin which do not appear to be involved in the regulatory mechanism. Thus the repeating structural unit of the regulated actin filament is A_7TmTn.

The complexity of the structure means that, unlike classical cooperative globular systems, it is not possible to define limits of cooperativity in either filament. Potential sources of cooperativity are along the thin filament, within and between A_7TmTn structural units, along the thick filament, between crowns of myosins and between myosin heads and the A_7TmTn units. However, the regular geometry does define some limits to the cooperativity. The stoichiometry of the protein complexes within each filament is fixed, as is the ratio of actin to myosin sites. In rigor, all of the myosin heads can bind to actin with a maximum of 67% of actin occupied by myosin in the overlap zone. Under contracting conditions (in the presence of ATP and Ca^{2+}) the degree of interaction between myosin heads and actin remains undefined (see below).

A discussion of the cooperativity of the complete system is beyond the scope of the current overview and we will therefore concentrate on the thin filament and its interaction with myosin heads. We will attempt to reduce the problem to the smallest cooperative unit and then consider to what extent the complete system can be rebuilt from these simple units.

4 The Three Thin Filament States from Solution Studies

4.1 Actin-S1 ATPase: Two Activity States

ATPase studies of myosin subfragment 1 (S1) interacting with the complete thin filament demonstrates features of the complex cooperativity seen in the skinned fiber measurements. The advantage of solution studies is that the concentrations of actin, S1 or ATP +/– Ca^{2+} can be varied, or the Ca^{2+} concentration can be varied at different protein concentrations. In the absence of actin, the S1 Mg^{2+}-ATPase is very low with a k_{cat} of 0.05 s^{-1}. The presence of pure actin filaments increases the ATPase up to 200-fold, hence our proposal that actin is the catalyst (Lehrer and Geeves 1998). In the presence of Tm, the plot of ATPase vs [S1] becomes cooperative and in the additional presence of Tn, the ATPase is dependent on Ca^{2+} (Fig. 3). At low [S1], the ATPase of the regulated actin filament (ATmTn) is inhibited compared with pure actin in both the presence and absence of Ca^{2+} but the inhibition is greater when Ca^{2+} is absent. As the [S1] is increased, the ATPase increases sigmoidally both in the presence and absence of Ca^{2+} above the S1 alone values, but the increase is greater when Ca^{2+} is present. This behavior is compatible with a model in which the S1 can cooperatively activate the regulated actin and Ca^{2+} acts as a classical allosteric activator of the thin filament activity (Lehrer and Geeves 1998). In agreement with the properties of allosteric effectors, the thin filament activity is *not* appreciably switched on by Ca^{2+} per se, but mainly influences the concentration at which the S1 is able to turn on the system. Strongly bound myosin heads, the product of the ATPase,

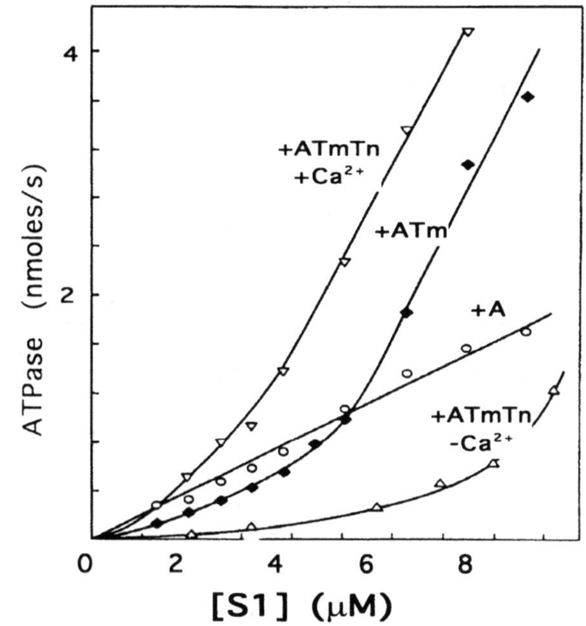

Fig. 3. ATPase vs [S1] for actin, actinTm, actinTmTn +/– Ca^{2+}. (After Lehrer and Morris 1982)

switch on the ATPase (Lehrer and Geeves 1998). This was observed in early ATPase studies that used NEM-S1, a modified S1 which binds strongly to the thin filament and is not dissociated by ATP (Pemrick and Weber 1976). The binding of only a few NEM-heads can turn on the ATPase and fewer heads are required for ATmTn in the presence of Ca^{2+}, compared with the absence of Ca^{2+} (Fig. 4). These results corroborated the pioneering studies in A. Weber's lab which showed activation of acto-S1 ATPase by rigor heads at low ATP concentrations in the presence of Tm and TmTn (Bremel et al. 1972) and the reciprocal relationship between myosin head and Ca^{2+} binding to the thin filament (Bremel and Weber 1972; Murray et al. 1975). Thus Ca^{2+}-induced cooperativity is a product of both Ca^{2+} and S1 binding to the regulated actin filament. During the ATPase reaction, the nature of the S1-nucleotide complex interacting with actin will change depending upon the level of activation. However, the fraction of the actin sites occupied by strongly bound S1 is ill defined. The system is much easier to study in solution when equilibrium binding of S1 to actin is considered.

4.2 Equilibrium Binding of S1 to Actin Filaments

Sedimentation and fluorescence methods have been used to estimate the fraction occupancy of actin filaments with S1 and S1-nucleotide complexes, as the S1 concentration is varied (Greene and Eisenberg 1980; Ishii and Lehrer 1985;

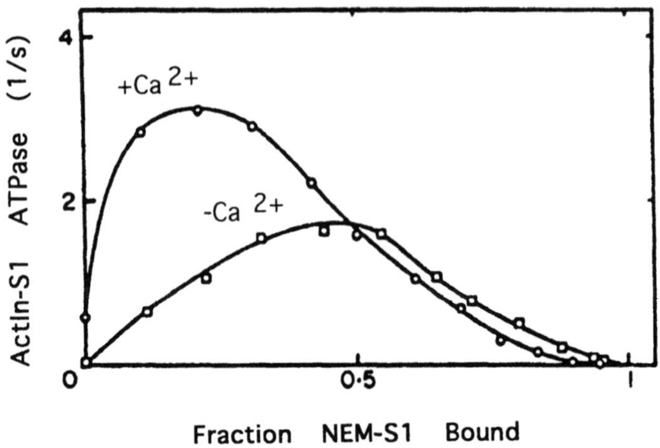

Fig. 4. The switching on of actin-S1 ATPase of regulated actin (actinTmTn) by the binding of low ratios of NEM-S1 in the absence and presence of Ca^{2+}. (After Nagashima and Asakura 1982)

Geeves and Halsall 1987). The binding of S1 to actin is non-cooperative in the absence of TmTn, indicating that the S1 binds to each actin site on the filament as if it were an isolated monomer with no interaction between actin sites. However, for ATm and ATmTn (+/−Ca^{2+}) the binding is sigmoidal, i.e., Tm alone can induce cooperativity of S1 binding as it does for the S1-dependence of actin-activated ATPase. Thus, at low [S1] the binding is inhibited, but after a fraction of the actin sites become occupied the system changes state (turns on) to a form which binds S1 more readily. As for the MWC classical cooperative system, four parameters are required to define the binding isotherm. These are; K_T, the equilibrium between the two states of the filament (on/off) in the absence of S1; n, the apparent size of the cooperative unit which changes state, K_1, the affinity of S1 for the off-state of the filament; K_2, the ratio of affinities of S1 for the two filament states (Geeves and Halsall 1987). For simple globular proteins the maximum size of the cooperative unit is defined as the structural unit, i.e., the number of subunits. In the case of the thin filament, the cooperative unit could be the repeating structural unit A_7 TmTn if it can change state independently. Thus, n can be <7 if Tm is very flexible, or >7 if the structural unit can communicate with its neighbor (see Sect. 6.3 below).

Two approaches were taken in fitting the cooperative binding isotherms; both using the assumption that $n = 7$. Hill et al. (1980) derived a model which included an extra cooperative parameter to allow the state of the two adjacent units to influence the probability of a change of state. The approach of Geeves and Halsall (1987) and McKillop and Geeves (1993) was a refinement of the Hill et al. model. Greene (1982) reported that the binding constants depended on the nucleotide bound to S1. Those constants were related to values of K_2, the value of the equilibrium constant for the isomerization from an A- to R-state in the two-step binding mechanism of S1 to actin (Geeves et al. 1984; Geeves 1991).

$$A + M \leftrightarrow A\text{-}M(A) \leftrightarrow A.M(\mathbf{R})$$
$$\quad K_1 \qquad\qquad K_2$$

It was proposed that in the closed state of the filament S1 could bind to form the A-state, but isomerization to the R-state was prevented by Tm. However, in the open state, S1 could form the R-state, trapping Tm in the open-state. Thus, the A to R transition of actoS1 and the closed-open transition form two coupled equilibria. Another feature of this model is that K_T and K_2 are independent of each other but K_T is Ca^{2+} dependent (K_T^+ and K_T^-, + and $-Ca^{2+}$, respectively) and K_2 is nucleotide dependent (Fig. 5).

Binding studies provided values to fit the Hill et al. and McKillop and Geeves models. These solution studies therefore define a cooperative binding model which is compatible with the ATPase data in that the thin filament is predominantly in the closed state and it is cooperatively switched to the open state by strongly bound myosin heads. Ca^{2+} does not turn on the thin filament, per se, but is simply an allosteric effector which allows S1 to switch on the system at lower concentrations. In muscle fibers the same model would apply, provided that the local myosin head concentration is within the relevant range (see Sect. 7 below).

4.3 Kinetics of S1 Binding to Actin Filaments: Three S1-Binding States

Evidence for a third state of the thin filament came from kinetic studies of the rate of S1 binding to actin (McKillop and Geeves 1993) which repeated earlier studies of Trybus and Taylor (1980) and integrated the ideas from both equilibrium and kinetic studies into a single scheme.

If the rate of S1 binding to pure actin filaments is followed in a stopped flow spectrophotometer and if the reaction is measured under pseudo-first order conditions (i.e., either actin or S1 in large excess over the other component) a simple exponential binding reaction is expected for actin (A) and S1 (M) as follows:

$$A + M \leftrightarrow A.M$$

for [M]>>[A] and an essentially irreversible process,

$$d[M]/dt = -[A][M].k$$

On integration this predicts an exponential decrease in [A].

$$[A]_t = [A]_0 \exp - k_{obs} t,$$

where

$$k_{obs} = [M]_{total} k$$

For [A]>>[M] a similar equation gives:

$$[M]_t = [M]_0 \exp - k_{obs} t,$$

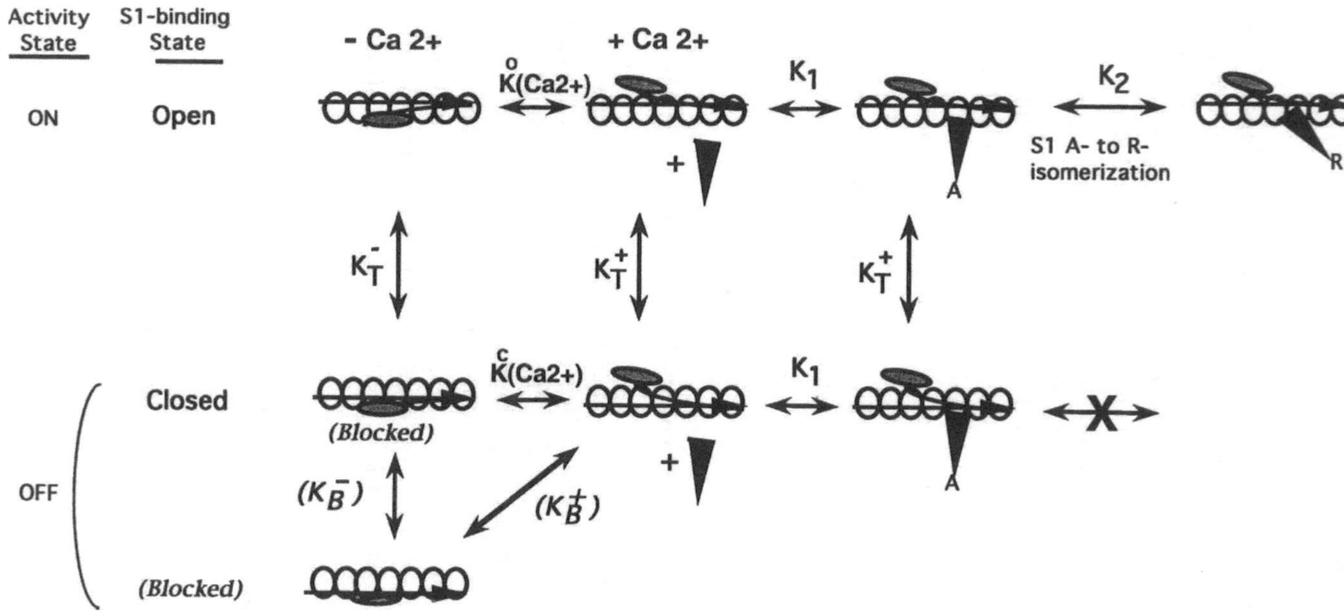

Fig. 5. Schematic diagram of the relationship of the three S1-binding states of actinTmTn, with the activity states and the A- and R-states of the S1-actin complex. Two interpretations of the blocked state are given: *bottom* an additional Tm position in equilibrium with closed state which blocks initial S1 binding; *middle* no additional Tm position but blocking by Tn exists in the closed state in the absence of Ca^{2+}. Equilibrium constants labeled with – and + are for the absence and presence of Ca^{2+}, respectively

where

$$k_{obs} = [A]_{total}.k$$

The equations accurately described the kinetics which gave the same k_{obs} for actin, ATm and ATmTn in the presence of Ca^{2+}. Thus, the kinetics indicate that all of the actin sites are available to the S1 in the presence of Ca^{2+}. In contrast, the equilibrium binding studies indicated that 80% of the actin sites are in the off-state.

In the absence of Ca^{2+} when the reaction is followed with actin in excess to S1 (i.e., only very few actin sites become occupied by S1) an exponential was also observed but k_{obs} was a factor of 3 lower than in the absence of Ca^{2+}. Thus, in terms of the above equations, either k or [A] was lower because a fraction of the actin sites was unavailable. In simple terms, if there are two states of the filament,

$$A(1) \leftrightarrow A(2) + M \leftrightarrow A(2).M$$

$$d[M]/dt = [A(2)][M].k$$

Assuming that the two forms of actin are in rapid equilibrium, the fraction of the total actin in the form A(2) is $K_B/(1+K_B)$, where $K_B = A(2)/A(1)$, hence:

$$d[M]/dt = [K_B/(1+K_B)][A]_{total}[M]k,$$

and

$$k_{obs} = [K_B/(1+K_B)][A]_{total}k$$

Since k_{obs} ($-Ca^{2+}$) is 1/3 of k_{obs} ($+Ca^{2+}$), the data can be explained by 30% of the actin monomers occupying the A(2) state or by a threefold smaller k. By varying the nucleotide bound to S1 (which changes k without affecting [A]) the value of k_{obs} in the absence of Ca^{2+} was always reduced to 30% of the value in the presence of Ca^{2+} for a range of values of k. Thus, the removal of Ca^{2+} affects K_B, not k. This suggests that the availability of actin sites is Ca^{2+}-dependent, but the fraction of sites available is very different when assayed kinetically or by equilibrium titrations and ATPase assays. Both methods show that all actin sites are available in the absence of TmTn. In the presence of TmTn and Ca^{2+}, the kinetic experiment showed that all actin sites are available but the equilibrium titrations and the ATPase data suggest only 20% available (a major discrepancy). In the absence of Ca^{2+}, the two methods also gave very different estimates: kinetically, 30% of the sites are available, whereas with the other methods <5% are available. The only simple resolution to this discrepancy was to propose that the two experimental methods measure different steps in binding. The sites which are available to bind S1 in a kinetic experiment are not the same as those which were able to bind S1 strongly in the titration. The three-state model to account for the new kinetic information is outlined here, where (off) and (on) refer to ATPase activity:

McKillop and Geeves three-state model (1993)

State 1	Blocked (off)	No productive S1-binding
State 2	Closed (off)	Binding to form the attached or A-state but no activation of the ATPase or force generation
State 3	Open (on)	Binding to form the rigor-like complex or R-state, accelerate product release and generate force

In the next Section we will consider the relationship of the blocked state to the $-Ca2^{+}$ state.

5 Relationship of Structural States to Solution States

Evidence for a three-state regulatory system has been obtained from X-ray studies on intact muscle (Holmes 1995) and EM studies on reconstituted thin filaments (Vibert et al. 1997; for a review see Squire and Morris 1998). Both methods appear to detect Ca^{2+} and S1-induced changes in structure. EM data suggest that Tm interacts with a large region of the myosin binding site on actin in the absence of Ca^{2+}, but that addition of Ca^{2+} moves Tm but still leaves it in a position to sterically hinder the formation of the myosin rigor-like complex. In the rigor complex the Tm is moved further away from the myosin binding site. This interpretation agrees well with the modeling of the thin filament by Holmes and his colleagues (Holmes 1995). However, neither method (EM or X-ray) considers the possible influence of Tn reorganization on the interpreted changes (Ishii and Lehrer 1990; Lehrer 1994; Squire and Morris 1998). Thus, it is possible that the blocked-state is a $-Ca^{2+}$ substate of the closed or off state with Tn subunits arranged differently than in the presence of Ca^{2+} (Fig. 5), rather than another position of Tm, or a combination of both. However, with this caveat of possible Tn reorganization, the structural and solution methods generally agree on the biochemical characteristics of the three states that are involved (see above). However, in addition to information on the blocked state, more information is needed about the equilibrium positions of all of the subunits and their interconversion kinetics under a given set of conditions and the size of the unit which can change state.

5.1 Observation of States

5.1.1 Probes of Equilibrium Titrations and Kinetics

The fraction of available actin sites in the three states has been probed using changes in light scattering, labeled actin fluorescence or sedimentation of actin to measure the binding of S1. It is desirable that the probes are sensitive to only one of the transitions; blocked to closed, closed to open, so that only data on a particular perturbation are monitored. The pyrene excimer fluorescence probe on Tm has proved useful in monitoring the closed to open transition. Probes of Tn subunit conformational changes have been performed with fluorescence (Trybus and Taylor 1980; Greene 1986; Zot and Potter 1987) and spin probes (Szczesna and Fajer 1995) which detect changes in environment and mobility. Proximity information has been obtained from cross-linking studies (Tao et al. 1990) and fluorescence detected energy transfer (FRET; Moens et al. 1994; Luo et al. 1998; She et al. 1998). Fluorescence images from extrinsically added labeled S1 bound to fibrils in rigor have revealed that heads are preferentially bound in the myosin-actin overlap region of the sarcomere, even in the presence of Ca^{2+} (Swartz et al. 1996). This verified that Ca^{2+} is not sufficient nor necessary to turn on the thin filament system. A variety of fluorescence probes between Tm and actin in reconstituted thin filaments at different amino acid sequence positions have not revealed changes in distance associated with Ca^{2+} binding (Tao et al. 1990; Miki et al. 1998), raising questions about whether Tm moves to a new average position as interpreted by EM and X-ray studies, or whether the FRET method is sensitive to the movement (Bacchiocchi and Lehrer 1999). Clearly, further information is needed to understand why probes on Tm do not indicate Ca^{2+}-dependent movement in contrast to EM and X-ray methods.

5.1.2 X-Ray Diffraction

X-ray diffraction patterns have been obtained from bundles of muscle fibers or oriented gels of regulated actin filaments before and after activation, either by electrical stimulation of muscle fibers or by adding Ca^{2+} or S1 to permeabilized muscle fibers or gels (Squire and Morris 1998). The diffraction pattern changes are interpreted in terms of fraction of the total structural change for a given S1 or Ca^{2+} concentration. The interpretation of the structural change is not straightforward. The contribution of random vs cooperative/bunched binding is not clear, nor is it always simple to define the influence of disorder/order transitions on the modeling. A change in intensity of a specific layer line could have contributions from several sources. Furthermore, the absence of a high resolution structure of Tn leaves any contribution of a structural change in Tn undefined.

5.1.3 Electron Microscopy

EM visualizes the predominant fixed structure present under a defined set of conditions. Normally, thin filaments from a suspension which attach to a substrate are fixed with stain (uranyl acetate) or frozen in a hydrated medium. Thus, they are "fixed" in the environment determined by the solution conditions with the assumption that the fixing does not alter the average structure. High-resolution images require the averaging of a large number of selected structures of good order and preservation. Helical averaging results in a uniform Tm strand with loss of information of the contribution of Tn and possible disorder. Non-uniformity along the Tm molecule and the globular "joints" at the ends (Phillips et al. 1986) are averaged out. With these caveats, the method determines the average positions which Tm can take on the actin surface provided these structures form a large fraction of the population. No information about dynamics or the contribution of minority structures is obtained.

6 Size of the Cooperative Units

Transitions between the three distinct structural/functional states turn on or off a certain number of actin subunits. We can define the size of the apparent cooperative unit, analogous to the "stoichiometry" of a binding transition, as the number of actin subunits "exposed" by binding ligand to one subunit. It is an apparent cooperative unit because a single S1 may not switch on all of the actin subunits in a cooperative unit because the free energy of binding may be insufficient. For hemoglobin, the maximum size unit that could be affected in the T to R transition, 4, is clear from the tetrameric structure of the protein. However, the apparent cooperative unit, *n*, for oxygen binding defined from the Hill coefficient, is smaller than 4 and its value depends upon the ligand used, the presence of any allosteric effectors and the general experimental conditions. For the muscle thin filament, the repeating structural unit is A_7TmTn, but the apparent cooperative unit theoretically could be greater than seven actin subunits because of communication across the filament, through actins, or along the filament through adjacent A_7TmTn units via Tm. A further complication in defining the cooperative unit is that this need not be the same for the blocked to open as for the closed to open transitions. In the next section we will consider the information on the cooperative unit size for the different transitions induced by Ca^{2+} and S1 binding.

6.1 Cooperativity of Ca^{2+} Binding to Troponin

Skeletal TnC has two low-affinity Ca^{2+} binding sites which trigger the muscle system and the data on Ca^{2+} binding to isolated TnC or to TnC complexed with TnI, whole Tn, TmTn or thin filaments suggest a Hill coefficient of between 1.5 and 2 (Potter and Gergely 1975; Grabarek et al. 1983). A smaller Hill coefficient is observed for cardiac Tn systems which have a single regulatory Ca^{2+} binding site. These observations are compatible with cooperative Ca^{2+} binding within the TnC without longer-range cooperativity between TnCs along the filament. An interesting observation is that the affinity of Ca^{2+} for TnC does vary for the different protein complexes. For isolated TnC and isolated thin filaments

(A_7TmTn), the affinity is relatively low with $K_d = 2$–$5\,\mu M$, but is almost tenfold tighter for the TnIC complex, for whole Tn, TnTm and for isolated thin filaments decorated with S1 (Grabarek et al. 1983). It is possible, therefore, that the two values reflect two different conformations of TnC in the complexes. A high affinity for Ca^{2+} is observed in any complex in which there is a strong TnI-TnC interaction. In the absence of TnI or when the TnI is strongly complexed to actin (in whole thin filaments but not in S1 decorated thin filaments) TnC has a lower affinity for Ca^{2+}. Thus, TnI can be viewed as an allosteric effector of Ca^{2+} binding to TnC and reciprocally Ca^{2+} is an allosteric effector of the TnI-thin filament interaction.

These equilibrium measurements do not estimate whether more than one A_7TmTn unit on actin has been affected by Ca^{2+} binding to the thin filament. However, the fact that $n<2$ for Ca^{2+} binding to TnC in thin filaments suggests that any such effect is relatively small.

6.2 Cooperativity of Blocked to Closed Transition

The kinetic experiments which reported the fraction of available actin sites (see Sect. 4.3) was studied over a range of Ca^{2+} concentrations. The resulting data (Fig. 6A,B) show that the fraction of actin sites in the blocked state switches from 0 to 30% over the pCa range 4–8 (Head et al. 1995). A fit to the Hill equation gave a midpoint at pCa = 5.6 and a Hill coefficient of 1.8, similar to the values observed for Ca^{2+} binding to the thin filament (Potter and Gergely 1975). Similar results were obtained for cardiac Tn with a smaller Hill coefficient (Reiffert et al. 1996). Thus, an equilibrium exists between the blocked and closed conformations of the A_7TmTn unit with $K_B = 0.3$ in the absence of Ca^{2+}, increasing to a value >10 in the presence of Ca^{2+}. We can use the information about the Ca^{2+} binding to Tn and the Ca^{2+} change in K_B to construct a simple MCW model of the equilibria:

$$
\begin{array}{ccc}
\text{Blocked} & \leftrightarrow & \text{Closed} \\
 & K_B = 0.3 & \\
K_{Ca} \updownarrow & & \updownarrow K'_{Ca} \\
\text{Blocked.Ca} & \leftrightarrow & \text{Closed.Ca} \\
 & K'_B & \\
K_{Ca} \updownarrow & & \updownarrow K'_{Ca} \\
\text{Blocked.Ca}_2 & \leftrightarrow & \text{Closed.Ca}_2 \\
 & K''_B &
\end{array}
$$

If the affinity of Ca^{2+} to the blocked state is that measured for TnC and ATmTn ($K_{Ca} = 2$–$5\,\mu M$) and for the closed state is that measured for a TnIC

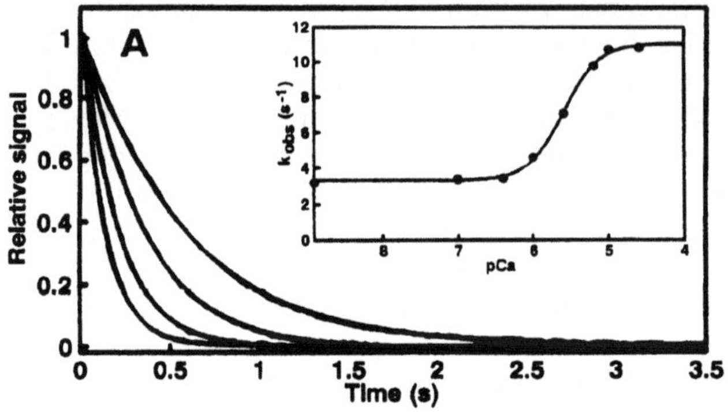

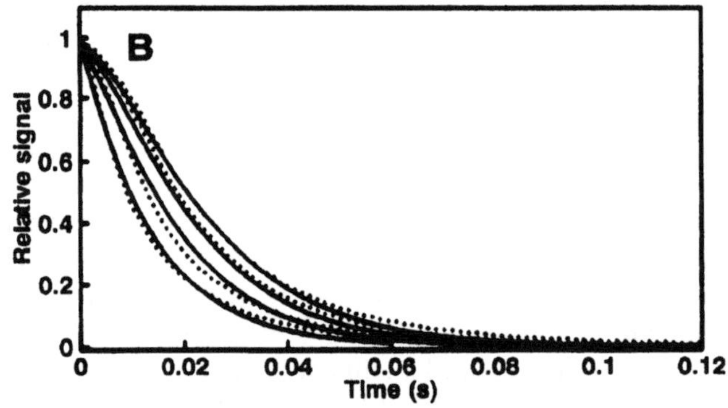

Fig. 6A,B. The Ca^{2+}-dependent binding kinetics of S1 to ATmTn. **A** Binding of 1-μM S1 to 5-μM actin, 1.4-μM TmTn at different pCa values, with fitted exponentials as *full lines*. The *insert* shows the pCa dependence of the rate constant. **B** Binding of 40-μM S1 to 1-μM actin, 0.28-μM TmTn at the above pCa values. (After Head et al. 1995)

complex or S1ATmTn (K'_{Ca} = 0.2–0.5 μM; see above), then from the relationship $K_B.K'_{Ca} = K'_B.K_{Ca}$ we can calculate that K'_B is approximately $10 \times 0.3 = 3$ and $K''_B = 30$. Thus, the binding of 0, 1 and 2 Ca^{2+} would switch the actin sites from 77 to 25 to 3% blocked, respectively. For a smaller difference between the two Ca^{2+} affinities, the switching is less complete; i.e., for a factor of 5 the fraction blocked would be 77, 40 and 11% for 0, 1 and 2 bound Ca^{2+}, respectively. This model shows that binding of one Ca^{2+} to skeletal Tn would cause a significant change in K_B. However, the cooperative binding of Ca^{2+} to Tn means that a Tn with a single Ca^{2+} bound would be very rare. The available evidence is that K_T, the equilibrium constant for the closed/open state, is much less affected by Ca^{2+} (three- to fivefold) and is always <1, so Ca^{2+} will have a smaller effect on the closed/open ratio, and the system will remain in the off ATPase activity state independent of Ca^{2+}.

The above argument does not reveal how many actin sites are switched on in the blocked to closed transition induced by Ca^{2+} binding. It can be assumed to be seven actin subunits if there is no Tn-Tn interaction. This was tested by adding an excess of S1 to filaments in the absence of Ca^{2+} and recording the kinetic transient (Fig. 6B). The fraction of actin sites in the two states (blocked and closed) at the start of the transient is defined by K_B at each Ca^{2+} (Fig. 6 A). The intrinsic rate constant for S1 binding is defined by the exponential at high Ca^{2+}. Thus, the only unknown in the model is *n*, the number of actin sites turned on by each S1 bound and therefore the sigmoid binding transients can be fitted by varying only one parameter, *n*. The data were shown to be compatible with $n = 7$ for the blocked to closed transition. That is, the switch from blocked to closed involves the switching of a single A_7TmTn unit. However, it should be pointed out that the model may not be very sensitive to large values of *n* but it is certainly not smaller than seven and the data requires no value larger than seven.

6.3 Cooperativity of Closed to Open Transition

The apparent cooperative unit size for the closed to open-transition has been measured in three ways. Ishii and Lehrer (1985) demonstrated that the excimer fluorescence of a pyrene label attached to Cys-190 of Tm (Tm*) signaled the closed to open transition of ATm and ATnTm. Thus, by combining light scattering signals (which measure the fractional occupancy of actin sites by S1) with the Tm* fluorescence (which measures closed to open), the average occupancy of actin with S1 required to change the Tm fluorescence was estimated in both equilibrium binding (Ishii and Lehrer 1985; Ishii and Lehrer 1990) and kinetic experiments (Geeves and Lehrer 1994). The kinetic experiments were performed in both the direction of S1 binding to actin and the ATP induced dissociation of S1 from the actin filament. These results showed that for skeletal muscle proteins a single S1 binding to one out of six to seven actins in A.Tm filament was sufficient to induce the change in Tm* fluorescence, i.e., $n = 6$-7. However, addition of Tn reduced the occupancy required to 1 in 12 sites, much larger than the structural unit. Thus, significant communication must exist between adjacent structural units. Further exploration of the reason for the larger apparent cooperative unit revealed that it was due to increased TmTm contacts, since the large cooperative unit only required the TnT_1 fragment of Tn which binds in the TmTm overlap region. The presence of $TnICT_2$, although conferring Ca^{2+} sensitivity on the S1-actin interaction, had no effect on the apparent cooperative unit size (Schaertl et al. 1995). Furthermore, use of gizzard Tm or fibroblast Tm, neither of which normally bind Tn and have stronger TmTm contacts, resulted in a cooperative unit size of 10–12 (Lehrer et al. 1997). Thus, the large cooperative unit size is a property of strong Tm-Tm contacts and the size of the cooperative unit does not relate simply to the length of Tm. This point is emphasized by the fact that gizzard and fibroblast Tm have similar cooperative unit sizes even though the fibroblast binds to only six actin sites compared with the seven of skeletal and smooth Tm. This observation is supported by recent results on the even shorter Tm from yeast (Maytum et al. 1998).

Thus, both kinetic and equilibrium measurements using pyrene-Tm agree on the apparent cooperative unit size for Tm. A third equilibrium method has given further support to this observation. A significant improvement in the sensitivity of equilibrium titrations of pyr-actin with S1 allowed titrations to be performed at actin concentrations as low as 50 nM, much lower than for the earlier titrations (Maytum et al. 1999). Under these conditions the titration observed is sensitive to the size of the cooperative unit as well as the other equilibrium constants. These titrations gave values of the cooperative unit ATm = 6–7 and for ATmTn = 12 in the presence of Ca^{2+}, in agreement with the earlier work. Surprisingly, in the absence of Ca^{2+}, the cooperative unit was equal to seven, suggesting that the TnI-A interaction, which repeats every seven actins, is strong enough to limit the longer-range cooperativity along the filament. The larger value for n obtained in the absence of Ca^{2+} for the ATP-induced S1 dissociation experiments (Geeves and Lehrer 1994) monitored the rate of the open to closed transition, where TnI binding to actin did not apparently influence the data.

EM studies addressed the question of how far beyond an S1 bound to a reconstituted thin filament does the Tm retain "memory" of the S1-induced local structure. Partially S1 decorated thin filaments were used. By noting patches of decorated and undecorated areas, the reconstructions showed that Tm moved between two positions (open to closed). They obtained values of about 300 Å–500 Å, or about ten Tm molecules (Vibert et al. 1997). These values, which are much greater than about the length of two Tm molecules which we obtained, can be taken as an upper limit of n, in view of the fact that a few S1s in the undecorated area may not be seen, which would keep the Tm in the open position.

These results lead to the question of what is the physical meaning of the size of a cooperative unit greater than that of the structural unit. We have discussed two interpretations (Lehrer et al. 1997; Maytum et al. 1999): one involves the model of Hill et al. where the state of a single A_7TmTn unit communicates with adjacent structural units via Tm end-to-end contacts. However, this would require major changes in the Tm-Tm contacts when structural units change state. The alternative, which we favor, is that when Tm-Tm contacts are strong, the Tm forms a continuous dynamic filament along the length of the actin filament (Fig. 7) with the apparent cooperative unit size a measure of the stiffness or the persistence length of the Tm filament on actin.

7 Relationship of Solution States to Fibers

The thin filament is thus a complex assembly of proteins which can exist in a number of structural and functional states, the observation of which depends on the methods and the conditions under which it is examined. The principal

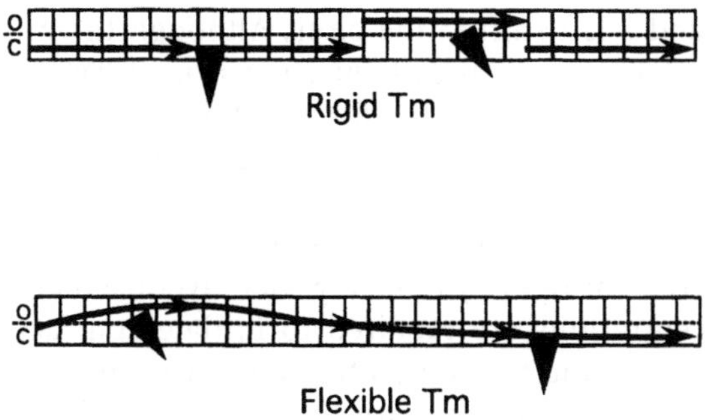

Fig. 7. Two models for tropomyosin trapped in the open state. (After Lehrer et al. 1997; Maytum et al. 1999)

effectors are Ca^{2+} and R-type myosin heads bound to the thin filament. We have shown that a self-consistent model can be produced from solution studies of the isolated components at equilibrium or kinetically during the approach to equilibrium. Unfortunately, we are still missing several key pieces of information to be able to describe the complete intact system. The most important missing piece is the number of cross-bridges strongly attached to actin during isometric contraction and/or the lifetime of an individual cross-bridge force-holding state. These numbers are still hotly debated among muscle physiologists with numbers of R-type bound cross-bridges ranging from 50 to 10% (Howard 1997; Huxley 1998) with recent studies favoring the lower number.

The above model can give a simple explanation of the force-pCa curves as the result of an increase in Ca^{2+} which reduces the occupancy of the blocked state leading to a recruitment of cross-bridges. The cross-bridges then cooperatively activate the filament leading to additional recruitment and hence the large Hill coefficient. Such a mechanism may cause non-symmetrical force-pCa curves (Fig. 1). However, the extent to which the plots deviate from symmetry is still the subject of debate (Moss 1992) and may be due to mechanical artifacts and non-homogeneity of the activation process throughout the muscle fibers as well as problems with calculations of $[Ca^{2+}]$ in EGTA buffer systems.

Because of the nature of the allosteric effect of Ca^{2+} on K_B and K_T and the coupling with K_2, we expect changes in the numbers of heads contributing to the force to change the Ca^{2+} sensitivity as observed (Hoffman and Fuchs 1987a,b; Gordon et al. 1988; Zot and Potter 1989; Millar and Homsher 1990; McDonald et al. 1995). Thus, the addition of any effector which acts on the myosin to alter the isometric force by changing the number of bound cross-bridges is expected to change the Ca^{2+} sensitivity, i.e., they are coupled. For example, Pi, acts as a competitive product inhibitor by reducing the number of force bearing cross-bridges and reducing isometric force, thereby producing a right shift of the force-pCa curve and decreasing the Hill coefficient (Millar and Homsher 1990). ADP acts in the opposite way to increase isometric force and a left shift in the plots. Although the primary site of action of phosphorylation of myosin light chains or of Tn components may be the alteration of the lifetime of the force-holding cross-bridges or changing the Ca^{2+} sensitivity, due to the coupling, effects on both force levels and regulation may be produced.

Allowing fibers to shorten can reduce the average number of strongly bound cross-bridges by reducing their average lifetime. In this case, higher Ca^{2+} concentrations are required to initiate contraction (rightward shift of the pCa curve) and the contraction becomes almost all or none, i.e., a very sharp transition and an increase in the Hill coefficient. But since the tip of the thick filament moves into an area of the inactive actin filament, the extent of this opposing the Ca^{2+} sensitivity is difficult to predict.

Even without an estimate of the number of cross-bridges involved in an isometric contraction, the above model of regulation can help define the limits of the number of cross-bridges involved. Since Ca^{2+} is limited to being an allosteric effector of myosin binding it will only begin to turn on contraction when there is a modest range of actin saturation with cross-bridges. If a single rigor-like cross-bridge within every 12 actin sites (the cooperative unit size) is sufficient to keep Tm in the on-state, then if half of the actin sites are occupied during a contraction, removal of Ca^{2+} will be ineffective at switching off the system. Equally, if the maximum occupancy of actin sites is 1 in 20 (5%) then Ca^{2+} will not provide enough bound cross-bridges to switch on the system. Note that here we are talking about the fraction of actin sites occupied, which is different from the fraction of myosin heads attached, as there are more actin sites than myosin heads in the overlap zone. In a mammalian skeletal muscle, the ratio of actin to myosin heads in the overlap zone is 3:2 so that the occupancy of actin sites in rigor is 67%. Assuming that at any moment only one of the two heads of each myosin interacts with actin in the presence of ATP, the maximal occupancy of actin sites is reduced to 33% (corresponding to 50% of myosin heads). The threshold for cooperative activation by myosin heads is 1 in 12 actin sites (8% of actin sites, or 12% of myosin heads). Ca^{2+} must be able to increase the number of cross-bridges across this 8% occupancy threshold for the system to turn on or off cooperatively in response to changes in Ca^{2+} concentration. Detailed modeling of this system is beyond the scope of the current review but an appreciation of the scale of the problem can be obtained in the following, with the use of the cooperative binding equation of McKillop and Geeves (1993).

In Fig. 8, a plot of the fraction of R-states (cross-bridges strongly bound to actin) as a function of apparent K_2 in the absence and presence of Ca^{2+}. K_2, the effective ratio of force holding R-states to weakly-bound A-states is related to the duty ratio as shown. This plot is meant to be a simple approximation to an isometrically contracting fiber to make the major point that the system is only Ca^{2+}-sensitive over a narrow range of values of K_2 from about 0.3 to about 3 (fractional actin occupancy of R-states from 0.02 to 0.5 for actin or 0.03 to 0.7 for myosin). Thus, for the Ca^{2+} switch to be effective, K_2 should be close to 1,

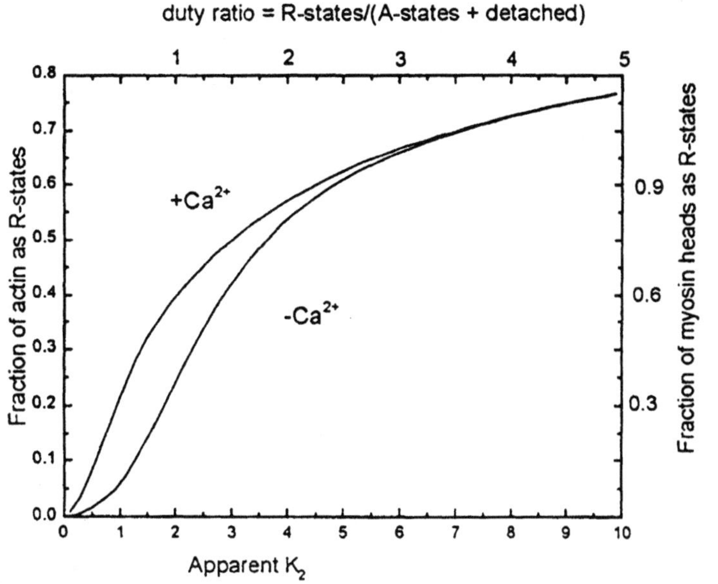

Fig. 8. Simulation of the fractional occupancy of actin by R-states (force holding cross-bridges) as a function of effective K_2 (cross-bridge isomerization constant) in the absence and presence of Ca^{2+}. The fraction of bound actin was calculated from equations for the cooperative binding of myosin heads to actin (McKillop and Geeves 1993). $K_1[M]$ was set equal to 1 (equal probability of actin unoccupied or occupied with A-state heads). K_2 defines the ratio of R- to A-states and is the average lifetime of the force holding R-state for cycling cross-bridges, and is related to the duty ratio [R-states/(A-states + detached heads)] as indicated. The fraction of myosin heads bound as R-states (1.5 of the number of actin subunits in the overlap zone of the sarcomere) is also shown. Ca^{2+} only has a significant effect on the fraction of R-states for a narrow range of values of K_2 (0.3–3) and for <50% occupancy of R-states

allowing the removal of Ca^{2+} to change the occupancy of actin sites from 0.3 to 0.05 or less.

Other mechanical parameters in muscle fibers such as force redevelopment kinetics (Brenner 1988; Hancock et al. 1997; Iwamoto 1998) and sinusoidal oscillations (Kawai et al. 1981) show the effects of Ca^{2+} regulation, but detailed interpretation of such processes is difficult while we cannot define the starting or end state of the system.

Acknowledgements. The support of grants from NIH HL 22461 and AR 41637 is appreciated.

References

Bacchiocchi, Lehrer SS (2000) Multi-site fluorescence energy transfer shows Ca^{2+}-induced tropomyosin movement in reconstituted skeletal muscle thin filaments. Biophys J 78:A364

Brandt PW, Cox RN, Kawai M (1980) Can the binding of Ca^{2+} to two regulatory sites on troponin C determine the steep pCa/tension relationship of skeletal muscle? Proc Natl Acad Sci USA 77:4717–4720

Bremel RD, Weber A (1972) Cooperation within actin filament in vertebrate skeletal muscle. Nat New Biol 238:97–101

Bremel RD, Murray JM, Weber A (1972) Manifestations of cooperative behavior in the regulated actin filament during actin-activated ATP hydrolysis in the presence of calcium. Cold Spring Harbor Symp Quant Biol 37:267–275

Brenner B (1988) Effect of Ca^{2+} on cross-bridge turnover kinetics in skinned single rabbit psoas fibers: implications for regulation of muscle contraction. Proc Natl Acad Sci USA 85: 3265–3269

Geeves M A (1991) The dynamics of actin and myosin association and the crossbridge model of muscle contraction. Biochem J 274:1–14

Geeves MA, Halsall DJ (1987) Two step ligand binding and cooperativity. Biophys J 52:215–220

Geeves MA, Lehrer SS (1994) Dynamics of the muscle thin filament regulatory switch: the size of the cooperative unit. Biophys J 67:273–282

Geeves MA, Goody RS, Gutfreund, H (1984) The kinetics of acto-S1 interaction as a guide to a model for the crossbridge cycle. J Muscle Res Cell Motil 5:351–361

Gordon AM, Ridgway EB, Yates LD, Allen T (1988) Muscle cross-bridge attachment: effects on calcium binding and calcium activation. Adv Exp Med Biol 226:89–99

Grabarek Z, Grabarek J, Leavis P, Gergely J (1983) Cooperative binding to the Ca^{2+}-specific sites of troponin C in regulated actin and actomyosin. J Biol Chem 258:14098–14102

Greene L (1982) The effect of nucleotide on the binding of myosin subfragment 1 to regulated actin. J Biol Chem 257:13993–13999

Greene L (1986) Cooperative binding of myosin subfragment 1 to regulated actin as measured by fluorescence changes of troponin I modified with different fluorophors. J Biol Chem 26: 1279–1285

Greene L, Eisenberg E (1980) Cooperative binding of myosin subfragment 1 to the actin-tropomyosin-troponin complex. Proc Natl Acad Sci USA 77:2616–1620

Hancock WO, Huntsman LL, Gordon AM (1997) Models of calcium activation account for differences between skeletal and cardiac force redevelopment kinetics. J Musc Res and Cell Motil 18:671–681

Head JG, Ritchie MD, Geeves MA (1995) Characterization of the equilibrium between blocked and closed states of muscle thin filaments. Eur J Biochem 227:694–699

Hill T, Eisenberg E, Greene L (1980) Theoretical models for the cooperative equilibrium binding of myosin subfragment 1 to the actin-tropomyosin-troponin complex. Proc Natl Acad Sci USA 77:3186–3190

Hoffman PA, Fuchs F (1987a) Effect of length and cross-bridge attachment on Ca^{2+} binding to cardiac troponin C. Am J Physiol 253:C90–C96

Hoffman PA, Fuchs F (1987b) Evidence for a force-dependent component of calcium binding to cardiac troponin C. Am J Physiol 253:C541–C546

Hoffman PA, Greaser ML, Moss RL (1991) C-protein limits shortening velocity of rabbit skeletal muscle fibres at low levels of Ca^{2+} activation. J Physiol (Lond) 439:701–715

Holmes KC (1995) The actomyosin interaction and its control by tropomyosin. Biophys J 68:2s–7s

Howard J (1997) Molecular motors. Nature 389:561–167

Huxley AF (1998) How molecular motors work in muscle. Nature 391:239

Ishii Y, Lehrer SS (1985) Fluorescence studies of the conformation of pyrene-labeled tropomyosin. Effects of F-actin and myosin subfragment 1. Biochemistry 24:6631–6638

Ishii Y, Lehrer SS (1990) Excimer fluorescence of pyrenyliodoacetamide-labeled tropomyosin: a probe of the state of tropomyosin in reconstituted muscle thin filaments. Biochemistry 29:1160

Iwamoto H, (1998) Thin filament cooperativity as a major determinant of shortening velocity in skeletal muscle fibers. Biophys J 74:1452–1464

Kawai M, Cox RN, Brandt PW (1981) Effect of Ca^{2+} ion concentration on cross-bridge kinetics in rabbit psoas fibers. Biophys J 35:375–384

Lehrer SS (1994) the regulatory switch of the muscle thin filament: Ca^{2+} or myosin heads? J Musc Res Cell Motil 15:232–236

Lehrer SS, Geeves MA (1998) The muscle thin filament as a classical cooperative/allosteric regulatory system. J Mol Biol 277:1081–1089

Lehrer SS, Morris EP (1982) Dual effects of tropomyosin and tropomyosin-troponin on acto-myosin subfragment 1 ATPase. J Biol Chem 257:8073–8080

Lehrer SS, Golitsina NL, Geeves MA (1997) Actin-tropomyosin activation of myosin subfragment 1 ATPase and thin filament cooperativity. Biochemistry 36:13449–13454

Luo Y, Wu J-L, Gergely J, Tao T (1998) Localization of Cys 133 of rabbit skeletal troponin-I with respect to troponin-C by resonance energy transfer. Biophys J 74:3111–3119

Maytum R, Konrad M, Geeves MA (1998) Effect of N-terminal mutations on the properties of yeast tropomyosin. Biophys J 74:A53

Maytum R, Lehrer SS, Geeves MA (1999) Cooperativity and switching within the three-state model of muscle regulation. Biochemistry 38:1102–1110

McDonald KS, Field LJ, Parmacek MS, Soonpaa M, Leiden JM, Moss RL (1995) Length dependence of Ca^{2+} sensitivity of tension in mouse cardiac myocytes expressing skeletal troponin C. J Physiol 483:131–139

McKillop DFA, Geeves MA (1993) Regulation of the interaction between actin and myosin subfragment 1: evidence for three states of the thin filament. Biophys J 65:693–701

Miki M, Miura T, Sano K-I, Kimura H, Kondo H, Ishida H, Maeda Y (1998) Fluorescence resonance energy transfer between points on tropomyosin and actin in skeletal muscle thin filaments: does tropomyosin move? J Biochem 123:1104–1111

Millar NC, Homsher E (1990) The effect of phosphate and Ca on force generation in glycerinated rabbit skeletal muscle fibers. J Biol Chem 265:20234–20240

Moens PC, Yee DJ, dos Remedios CG (1994) Determination of the radial coordinate of Cys 374 in F-actin using fluorescence energy transfer. Biochemistry 33:13102–13108

Monod J, Wyman J, Changeux J-P (1965) On the nature of allosteric transitions. J Mol Biol 12:88

Moss RL (1992) Ca^{2+} regulation of mechanical properties of striated muscle. Circulation Res 70: 865–884

Murray JM, Weber A, Bremel RD (1975) Could cooperativity in the actin filament play a role in muscle contraction? In: Carafoli E (ed) Calcium transport in contraction and secretion. Elsevier/North-Holland, New York, pp 489–496

Nagashima H, Asakura S (1982) Studies on cooperative properties of tropomyosin-actin and tropomyosin-troponin-actin complexes by the use of N-ethyl maleimide-treated and untreated species of myosin subfragment 1. J Mol Biol 155:409–428

Pemrick S, Weber A (1976) Mechanism of inhibition of relaxation by N-ethyl maleimide treatment of myosin. Biochemistry 15:5193–5198

Phillips GN, Fillers JP, Cohen C (1986) Tropomyosin crystal structure and muscle regulation. J Mol Biol 192:111–131

Potter JD, Gergely J (1975) Troponin tropomyosin and actin interactions in the regulation of muscle contraction. Biochemistry 13:2697

Reiffert SU, Jacquel K, Heilmeyer LMGJ, Ritchie MD Geeves MA (1996) Bisphosphorylation of cardiac troponin I modulates the Ca-dependent binding of myosin subfragment 1 to reconstituted thin filaments. FEBS Lett 384:43–47

Schaertl S, Lehrer SS, Geeves MA (1995) Separation and characterization of the two functional regions of troponin involved in muscle thin filament regulation Biochemistry 34:15890–15894

She M, Xing J, Dong W J, Umeda P K, Cheung H C (1998) Calcium binding to the regulatory domain of skeletal muscle troponin C induces a highly constrained open conformation. J Mol Biol 281:445–452

Squire JM, Morris EP (1998) A new look at thin filament regulation in vertebrate skeletal muscle. FASEB J 12:761–771

Swartz D, Moss RL, Greaser ML (1996) Calcium alone does not fully activate the thin filament for S1 binding to rigor myofibrils. Biophys J 71:1891–1904

Szczesna D, Fajer PG (1995) The tropomyosin domain is flexible and disordered in reconstituted thin filaments. Biochemistry 34:3614–3620

Tao T, Gong B-J, Leavis PC (1990) Ca-induced movement of TnI relative to actin in skeletal muscle thin filaments Science 247:1339–1341

Trybus KM, Taylor EW (1980) Kinetics of the cooperative binding of subfragment 1 to regulated actin. Proc Natl Acad Sci USA 77:7209–7213

Vibert P, Craig R, Lehman W (1997) Steric-model for activation of muscle thin filaments. J Mol Biol 266:8–14

Zot AS, Potter JD (1987) Structural aspects of troponin-tropomyosin regulation of skeletal muscle contraction. Annu Rev Biophys Chem 16:535–559

Zot AS, Potter JD (1989) Reciprocal coupling between TnC and myosin crossbridge attachment. Biochemistry 28:6751–6756

Motility Assays of Calcium Regulation of Actin Filaments

M.A. LaMadrid, P.B. Chase, and A.M. Gordon[1]

1 Introduction

The exact details of how calcium regulates muscle contraction is still an important problem in the motility field. Actin filaments in muscle have additional proteins, troponins and tropomyosin, which regulate how myosin interacts with actin. Various chapters in this volume, by S. Lehrer and M. Geeves, R. Craig, M. Miki, T. Wakabayashi and L. Tobacman, as well as recent reviews (e.g., Farah and Reinach 1995; Tobacman 1996; Squire and Morris 1998; Gordon et al. 2000) describe the structural and biochemical changes accompanying calcium regulation at the actomyosin interface. X-ray diffraction and electron micrographic data show that calcium induces changes may be ascribed to movement of either tropomyosin (Tm) alone or to both troponin (Tn) and tropomyosin (Squire and Morris 1998). Biochemical studies (Lehrer and Morris 1982; Williams et al. 1988) show that binding of myosin to actin can be cooperative in the presence of troponin, tropomyosin and calcium. Skinned fiber studies show that the calcium activation at steady force appears cooperative (Brandt et al. 1984), but that the calcium dependence of the rate of force development can be explained with little cooperativity on the basis of control of attachment (Regnier et al. 1998). Does calcium merely control the number of cross-bridges that can bind to actin or does it also control the kinetics of the of the actomyosin interaction? How is this control implemented on the molecular level?

The in vitro motility assay is a powerful tool that can be used to answer these molecular questions on calcium regulation. In this assay, purified motor proteins are adsorbed on a glass surface, and the gliding of single fluorescently labeled filaments is observed by video microscopy. The details for myosin-specific motility studies have been published (e.g., Kron et al. 1991; Sellers et al. 1993).

For studies of calcium regulation, the ability to reconstitute the thin filament from regulatory proteins (from different sources and isoforms) in various combinations, and use different myosins makes the in vitro motility assay extremely powerful. Some of the other advantages include simpler quantitative analyses, removal of sarcomeric constraints, the ability to use small

[1] Dept. of Physiology and Biophysics, Dept. of Radiology, University of Washington, Seattle, Washington 98195, USA

Results and Problems in Cell Differentiation, Vol. 36
D.D. Thomas, C.G. dos Remedios (Eds.): Molecular Interactions of Actin

amounts of proteins and the ability to analyze the action of a small number of molecules and vary their numbers.

In recent years, some groups have started to use this technique to study calcium-regulated actin filaments. Studies with rabbit skeletal proteins (Honda and Asakura 1989; Harada et al. 1990; Honda et al. 1995, 1996, 1997), and with rat cardiac proteins (Sata et al. 1995c) have found that the speed of regulated actin filaments with reconstituted troponins and tropomyosins can go from zero to maximum in an apparently on-off manner, over a small range of [Ca^{2+}]. These studies have been mainly performed at low ionic strength, which may influence the interactions between the regulatory proteins (Gordon et al. 1973; Gulati 1983). On the other hand, other studies at more physiological ionic strengths using combined rabbit skeletal and bovine cardiac proteins (Homsher et al. 1996; Lin et al. 1996) or rabbit skeletal proteins (Fraser and Marston 1995; Gordon et al. 1997) found a more graded behavior. It was also observed that it is the presence of troponin which confers the in vitro calcium sensitivity as tropomyosin alone does not (Bing et al. 1997a; Gordon et al. 1998).

In addition to using motility assays with the wild-type reconstituted purified proteins, other studies have used the technique to evaluate the effects of using various recombinant proteins, modifications by drugs or other agents and to investigate Ca regulation of the thin filament. It has been found (Sata et al. 1995a,b) that the drugs pimobendan and MCI-154 can shift the [Ca^{2+}] concentration at which a cardiac thin filament can be activated, as they shift Ca sensitivity in cardiac muscle. A mutation in troponin-T that is related to familial hypertrophic cardiomyopathy (FHC) has been found to increase the gliding speed of thin filaments. Deletion of three of the seven units in tropomyosin has been found to decrease the gliding of thin filaments (Lin et al. 1996) in the presence of Ca, but addition of NEM S-1 can reverse this inhibition (Landis et al. 1997). This implies a role of strongly attached crossbridges in moving Tm. Further support for this comes from another study where it was found that adding chemically modified HMM (IASL-S1) to normal HMM in a motility buffer increases speeds at submaximal [Ca^{2+}] concentrations like pCa 7–8 (Bobkov et al. 1997).

More recently, attempts have been made to measure the force generated by myosin molecules acting on an actin filament in combination with the motility assay. One method of measuring force was to attach the actin filament to a glass microneedle and measure the deflection of the microneedle. Using this combination, the force and gliding speed vs pCa have been simultaneously measured using thin filaments with cardiac regulatory proteins (Lee et al. 1998). It was found that the force need not be at its maximum value for a filament to move at maximum speed. Another technique for measuring the force generated is to load the filament with pPDM-modified HMM (which binds weakly to actin) together with HMM adsorbed on the glass surface. With this technique it was shown that reconstituting cardiac Tn with the familial hypertrophic cardiomyopathy TnT mutant can result in an increase in the regulated thin filament's speed for maximal [Ca^{2+}] but with a decrease in force (Lee et al. 1997; Lin et al. 1996). This may help explain the hypertrophy observed in this condition.

In this chapter, we describe a recent application of the in vitro motility assay in our laboratory's attempt to understand how [Ca^{2+}] regulates muscle contraction. We previously showed that in rabbit skeletal muscle, [Ca^{2+}] grades the gliding speed of regulated thin filaments under physiological (140 mM) ionic strength conditions (Gordon et al. 1997). Further analysis of the gliding speed data was performed in an attempt to distinguish whether the calcium was affecting the number of motors that strongly bound to actin, whether it was affecting a kinetic step, or both. The calcium regulation of speed was measured as a function of thin filament length.

The data were then analyzed in the context of two limiting cases for unloaded shortening speed – the case when there is no significant drag (when the number of motors and the duty ratio would set the maximum gliding speed), and the case when drag is significant (when protein friction of various origins would limit the speed). In the case of insignificant drag, if the duty ratio of the myosin motor was unchanged by calcium, and if calcium was merely controlling the number of motors that can bind to actin, then short regulated filaments should move more slowly than long filaments at a low enough calcium concentration. In the case of significant drag, different forms of drag would have different dependences on filament length. If the drag is due to opposing effects from strongly bound crossbridges, then speed should not depend on the its length, since both force and drag would change proportionately.

At the [Ca^{2+}] studied, it was found that although short filaments had a higher tendency to stop, when they were moving, their speeds were not significantly different from the long filaments. The possible implications are that calcium may be changing the strong binding time, thereby changing the duty ratio and/or the drag on the actin filament, in addition to controlling the number of motors that can bind.

2 Materials and Methods

The experimental details for this study were identical to those in Gordon et al. (1997). Myosin, actin, tropomyosin and troponin were all obtained from rabbit back muscle. Regulated actin filaments were made by incubating rhodamine-phalloidin labeled F-actin with troponin and tropomyosin. The data analyzed for length dependence of the speed were obtained at 30 °C from motility assays with 0.25 mg/ml [HMM], 8–10 nM actin or regulated actin, 2 mM [MgATP], 0.14 M Γ/2, pH 7.0, anti-photobleaching and anti-photooxidation enzymes. Methyl cellulose (0.7%) was added to prevent actin filament diffusion from the surface. pCa was varied using appropriate amounts of EGTA and calcium propionate. Tn and Tm in 4:5 molar ratio were also added to the motility solution to ensure that there was no observable motility at pCa 9.2.

Speed data were obtained by videotaping six to eight fields of fluorescently labeled filaments for 40–60 s, then digitizing the images using a commercial package called ExpertVision(EV) by Motion Analysis Corporation (ExpertVision 1990), described in Homsher et al. (1992) and Sellers et al. (1993). The EV system greatly simplifies and speeds up the analysis of large numbers of filaments or of high contrast objects in general. This system has hardware that can simultaneously track many filaments, and comes with software that enables the user to calculate relevant parameters such as speed and speed distributions. (It should be noted that the more commonly used method of calculating the speed for a few operator-selected filaments obviously introduces bias into the choice.) The data were collected at 30 frames/s, then points were averaged to give effective frame grabbing speeds of 10, 5 or 2 frames/s, depending on the average speed of filaments. In addition to optimal magnification and pixel density, there is an appropriate optimal sampling rate for filaments moving at a particular speed – sampling too fast accentuates noise which can lead to overestimation of the speed, while sampling too slowly can artificially shorten a curved trajectory, giving an underestimate of the speed.

A filament is detected via hardware edge detection. Centroids are then calculated from the edges of each detected object in each video frame, and the speed of the filament is calculated from the trajectory of the time series of centroids. A filament must be unambiguously identified in (typically) 20 sequential frames to form a time series (path). Filaments of different length ranges can be separated by changing the parameters for identifying objects in the centroid calculation to require objects to have a minimum and maximum number of pixels to be included in a particular length group. The resulting files had filaments grouped into four subdivisions: L~1–3 μm, L~4–6 μm, L~7–12 μm, L>~12 μm. The lengths are approximations because the relationship between the object definition parameters and actual lengths are complicated by contrast issues and curvature of the filaments; however, it can be generally said that the average length in the different groups scales in the same way as the number of pixels needed to define a filament.

The object definition parameters for the centroid calculation were also chosen to minimize the possibility of a filament being classified in two groups. The apparent size of a filament can change during recording for various reasons such as variations in illumination level across a field, photobleaching, filament bending, pixel noise, and proximity to the screen edge. A filament's real size may also change due to fragmentation. Because of these apparent and real size changes, the total number of filament paths in all size groups was typically less than the total number of filament paths identified when the only size restriction was L>~1 μm. (This minimum filament size is set by the pixel resolution limit of 0.4 μm/pixel, noise and the centroid cutoff of 4×4 pixels.) Some filaments that were included in the initial analysis of all sizes combined were automatically excluded from the analysis with the restricted size bins because they did not unambiguously meet the required number of frames to make a path. This excluded number is estimated to be less than 10% of the total filaments analyzed. The effect of filaments entering or leaving the screen was quantified by observing the speed and standard deviation versus time for 50–75 short (L~1–3 μm) filaments with short paths, where the effects are maximized. The speed record for the section of their trajectories when they were either entering or leaving the screen did not have significant differences from the rest of the trajectory.

Speed is calculated from the trajectory traced out by the identified centroids in successive frames. In this analysis, speed is defined as total length of the trajectory divided by the total time. The speed data shown in the speed versus pCa plots are obtained by averaging the speeds of all the filaments in each length group. There was an original total of ~9000 trajectories analyzed, then subdivided into the different length groupings. Each data point represents the average of 30–440 trajectories. Standard errors range between 2 and 14% with a 7% average standard error.

3 Results

Figure 1A shows the unweighted average speed of different filament length groups for different [Ca^{2+}]. In all length groups, the speed increases in a nonlinear fashion with [Ca^{2+}]. However, there is an apparent difference in average speeds between the shortest filaments, L~1–3 μm, and the rest of the filaments, for all pCa's. There is no significant difference in the speeds of filaments longer than 3 μm.

Examination of the videos indicated that the shortest filaments displayed saltatory (stop/go) motion more frequently than longer filaments, especially with lowered [Ca^{2+}]. This observation was supported by the data in figure 1B which shows the fraction of moving filaments in the different length groups as a function of [Ca^{2+}]. This fraction was calculated by subtracting the background noise distribution, and counting the number of filaments left with speed higher than the noise generated speed. It is not clear if these apparently non-moving short filaments are truly immobile or are diffusing more slowly than can be observed. It is plausible that a minimum number of motors are required to start the motion of a filament and overcome whatever forces are keeping the filament immobile, or that strongly bound cross-bridges play a role in turning on the thin filament.

If shorter filaments move less frequently than longer ones, then their lower apparent speed would arise from this fact. We questioned whether they had lower speeds when they were moving. To estimate the speed in the moving portion of the filaments' trajectories, the apparent speed due to the background noise distribution of the imaging and EV system, measured by digitizing filaments in rigor, was subtracted from the average filament speed distribution. The recalculated speeds, which are the average speeds from the portions of the trajectories which truly represent motion, are shown in Fig. 2. Figure 2 shows

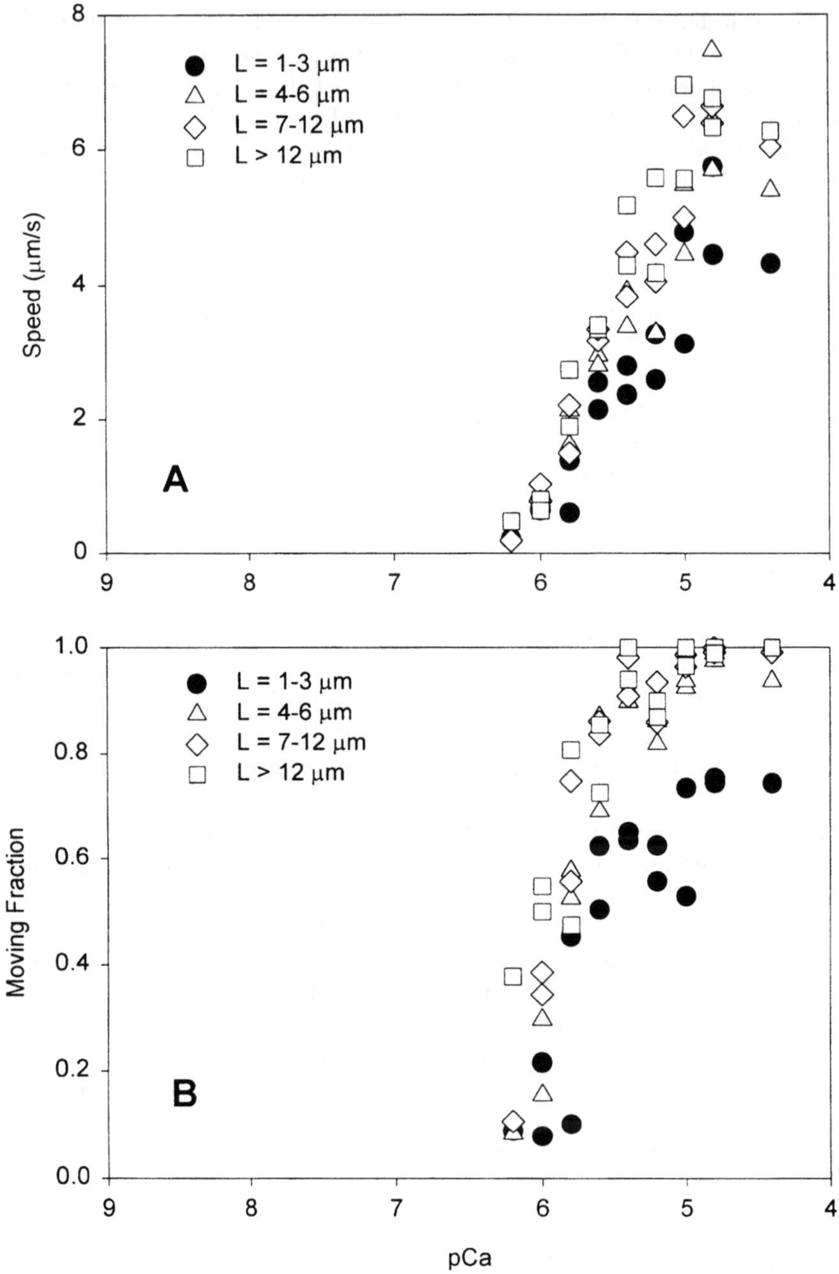

Fig. 1. **A** Average filament speed vs pCa for different filament length groups. The speed is the total distance traveled divided by the total time. **B** Fraction of moving filaments vs pCa. Note that the shortest filaments spend less time moving

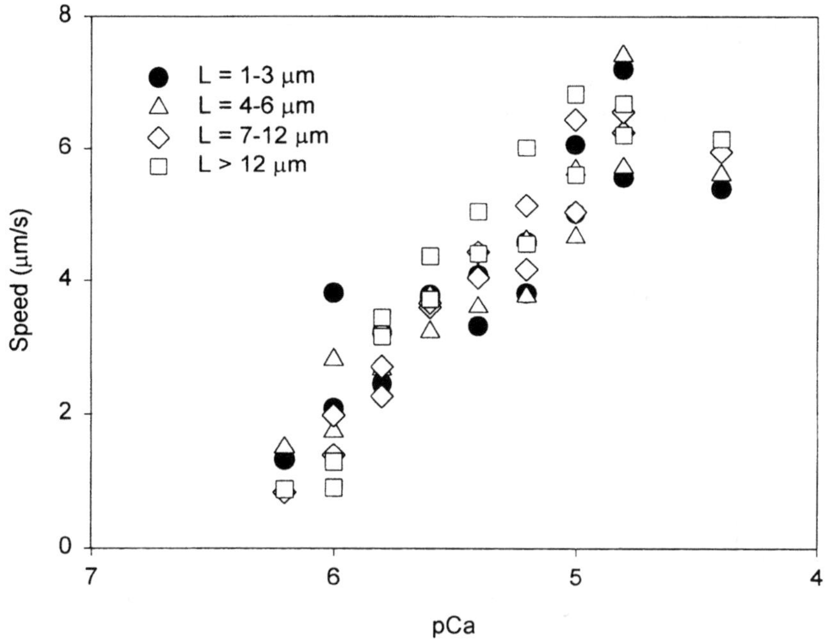

Fig. 2. Average filament speed vs pCa AFTER subtracting the non-moving portions of the thin filaments' trajectory. There is no significant difference in the speeds of the different length groups

that there are no significant differences in the average speeds of the different length groups *when the filaments are moving.*

Speed can also be defined for "smoothly" moving filaments by including only filaments whose trajectories satisfy standard deviation <0.5(mean; Homsher et al. 1992, 1996). This analysis of data in Fig. 1A shows no significant differences between the different length distributions.

4 Discussion

Analysis of the length dependence of speed shows two results that need to be explained: first, there is no significant difference in the speeds of MOVING regulated filaments ranging in length from ~1 to >12 μm. Second, filaments ~1- to 3-μm long spend less time moving, but move at essentially the same speeds as longer filaments when they do.

What gives rise to the observed [Ca^{2+}] dependence of in vitro motility speeds? Figure 3 shows that there is no change in speed with [Ca^{2+}] if the F-actin does not contain Tn and Tm, as observed by others (e.g., Homsher et al. 1996). This implies that [Ca^{2+}] does not modulate the function of HMM or unregulated actin in this system, as may be the case with other myosins, e.g.,

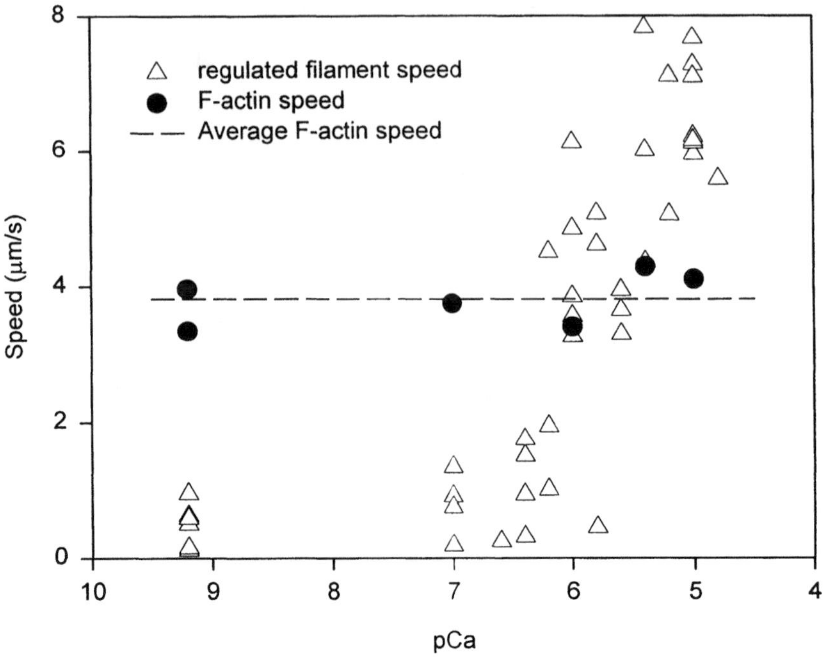

Fig. 3. Comparison of F-actin speed with regulated filament speed at different pCa's. F-actin speed has no calcium dependence in the range studied. Note that regulated filament speed at 5 pCa is significantly higher than F-actin speed. Data were obtained in 0.085 Γ/2 ionic strength and 0.4% methyl cellulose

scallop myosin (Xie et al. 1994; Szent-Györgyi 1975). In this system, $[Ca^{2+}]$ affects the actomyosin interaction via troponin and tropomyosin.

If the dependence on $[Ca^{2+}]$ is related to the number of motors that can bind and exert force on the filament, as the steric blocking model suggests (e.g., Vibert et al. 1997), there are two possible limits for the filament speed. One case is when drag is negligible and what limits speed is purely kinetic. The other case is when drag is not negligible and what limits the speed is the balance between the positive motor forces and the drag. We explore the implications of the data for both models.

As originally proposed (Harada et al. 1990; Uyeda et al. 1990) if there is no drag, and motors work independently, the in vitro motility speed is given by $V = (d/t_s)[1 - (1 - b)^N]$, where d is the myosin power stroke, t_s is the strong binding time, b is the duty ratio, which is t_s/t_{cycle}, N is the number of motors that are stroking, and t_{cycle} is the ATP cycling time. The duty ratio is the fraction of time that the myosin is producing force (Howard 1997). Since N is dependent on the length of a thin filament (as well as the density of motors on the surface), the drag-free case predicts that the filament speed should depend on the filament length, depending on the value of the duty ratio and motor density. At

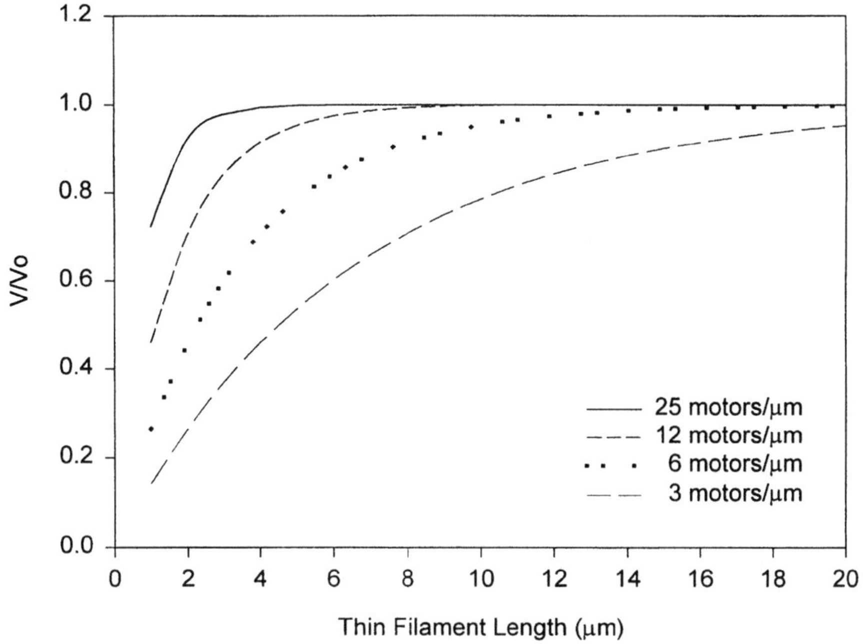

Fig. 4. Plot of $V/V_o = 1 - (1 - b)^N = 1 - (1 - b)^{\rho L}$ for different ρ's, where N is the number of attached motors, ρ is the number of motors/unit length, L is the filament length, $V_o = d/t_s$, and b is the duty ratio (0.05). Note that for the lower densities there is a strong length dependence

high enough duty ratio and density (e.g., >10% duty ratio, >12 motors/μm), the speed is length independent. Below a certain range of motor density and duty ratio, a length dependence should be observable. Figure 4 shows the predicted length dependence of the normalized speed for different motor densities and a duty ratio of 5% – a value close to that found in other in vitro motility work (e.g., Uyeda et al. 1990; Harris and Warshaw 1993). For example, for a motor density of 6 motors/μm, a 10-μm-long filament would move at close to maximum speed, while a 3-μm-long filament would only move at 70% of the top speed. In addition, evaluating the above equation for a maximal speed of ~8 μm/s implies a duty ratio of 1–2% or t_s~0.625 ms, if d~5 nm (Molloy et al. 1995b; Mehta et al. 1997) and t_{cycle} ~36–50 ms (Greene at al. 1987; Toyoshima et al. 1987), depending on the temperature (Brenner and Eisenberg 1986); the length dependence is easily observable for L~2–10 μm.

Figure 2 shows that the speeds of moving portions of the trajectories have no significant length dependence. This could imply that $[Ca^{2+}]$ does not merely affect the number of motors that can attach. On the other hand, this can also mean that there are always plenty of available motors that can bind and stroke productively.

How many motors can interact with actin productively? At 250 μg/ml, our ATPase assays indicate that the surface density is ~500 functional HMMs/μm^2.

The number of motors per unit length that can interact with the filament is equal to the motor density multiplied by the number that can reach the filament (Uyeda et al. 1990). Assuming a minimum width of 30 nm for the band of myosins that can reach the filament (Uyeda et al. 1990), and that only one head at a time can bind, a minimum number of 15 motors/μm can reach an actin filament. However, since an HMM complex can be 65 nm long (Bagshaw 1993), this can increase the number to ~90 motors/μm, since up to two rows of HMMs on either side of the filament can interact. Hence, at a motor density of ~500 HMMs/μm^2, from ~15–100/μm motors can simultaneously interact with a filament, depending on how the HMM is oriented, and assuming there are no steric constraints on the ability to bind.

How many of the ~15–100 motors/μm can actually bind AND perform a power stroke can be less than how many can bind. Recent studies (Sase et al. 1997) show that actin does not rotate very much during gliding (one rotation in >1 μm of translation). This implies that motors are not attaching in adjacent actin monomers, skipping many actin monomers ("running instead of walking" – Howard 1997) and are probably attaching at sites related to the helical period of the actin filament. A recent experiment (Molloy et al. 1995a) where a filament was bowed back and forth by a myosin head showed preferential attachments at ~40-nm intervals. Since our filaments have Tn and Tm, the number of sites a motor can bind to is severely constrained, i.e., in a 1-μm filament, perhaps only 25 motors, at most, can bind AND produce force at a time. If this is the case, then there should be a clear length dependence for a duty ratio <5% as [Ca^{2+}] is decreased.

No observed length dependence implies >5% duty ratio with this analysis, contrary to other in vitro motility results (Uyeda et al. 1990; Harris and Warshaw 1993). If the duty ratio were >5%, then the length dependence would be hard to detect for filaments longer than the 1.2-μm minimum for our apparatus. Shorter filaments would have to observed. Fiber studies imply a duty ratio of >10–20% during unloaded shortening (e.g., Julian and Morgan 1981; Ford et al. 1985; Griffiths et al. 1993). If this is the case, then there would be no observable regulated filament length dependence in the motility assay in the 1- to 20-μm length range. It is also possible that changing the [Ca^{2+}] changes the duty ratio. This possibility will be discussed later.

The other limiting case for the in vitro gliding speed assumes that there is significant drag. In the presence of drag, various authors (e.g., Tawada and Sekimoto 1991; Cuda et al. 1997) show that the speed is given by the balance between the positive displacement producing forces and the drag producing forces, i.e.,

$$v \sim \text{sum of positive forces}/\text{sum of drag coefficients},$$

where the sources of drag can include hydrodynamic drag, weak binding drag, and drag due to negatively strained motors, and it is assumed that the drag forces are all proportional to speed. It can be estimated that hydrodynamic drag on the actin filament at typical speeds and viscosity of the motility assay does not significantly affect this system. If generalized "protein friction" is the source of non-hydrodynamic drag (e.g., Tawada and Sekimoto 1991; Leibler and Huse 1993), then

$$v \sim [f_d P_d N_d/L]/[(kt_h/2)P_h(N_h/L)],$$

where f_d is the motor force, P_d is the probability of the motor being in a force-producing state, N_d/L is the number of motors per unit length that can interact with the filament, $kt_h/2$ is the drag coefficient for protein friction (k is the protein spring constant, t_h is the holding time, which is the time the protein is "dragging"), P_h is the probability of being in a "holding state", and N_h/L is the number of motors per unit length that can interact with filament in the "holding" state. Similar results have been obtained by other authors, where the details of the calculation may differ, but the physics is the same.

One possible source of protein friction is that due to so-called weakly bound cross-bridges (Schoenberg 1988). Two pieces of evidence argue against this possibility for this system. First, estimation of the amount of drag that can be produced by weak binding and using a weak binding time of ~0.1 ms (Schoenberg 1988) predicts in vitro speeds of >200-μm/s, which we do not observe. Second, increasing the ionic strength should significantly decrease the weak binding and increase the speed. Figure 5 shows the in vitro speed for three different ionic strengths, and no significant difference in speed is observed.

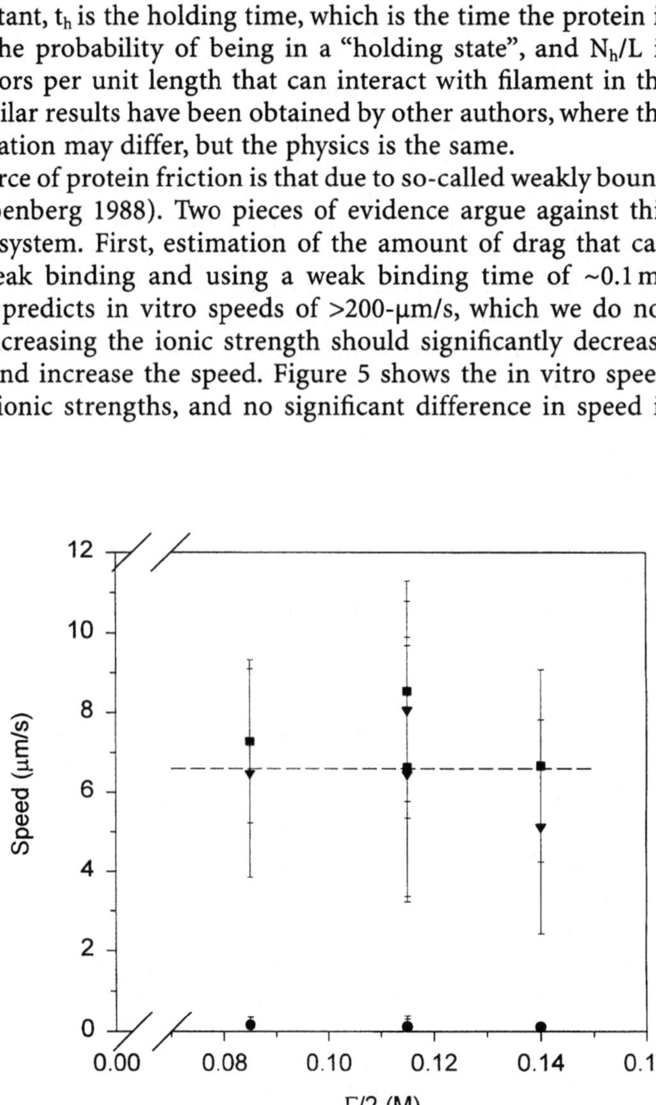

Fig. 5. Speed vs ionic strength. Friction due to weak binding is predicted to have a strong dependence on ionic strength

The drag due to strongly attached negatively strained crossbridges is a plausible source of drag which gives predictions consistent with the results above. Such crossbridges are commonly thought to limit unloaded shortening speed in fibers and intact sarcomeres. If so, then the corresponding speed would be approximated by $v \sim 2 f_d/kt_s$, where t_s equals t_h, which is the strong binding time, and N_s equals N_h, so number dependence cancels out. For a strong binding time of ~1–10 ms (Finer et al. 1994), $f_d \sim 1$–10 pN (Finer et al. 1994; Molloy et al. 1995a,b), k~1 pN/nm (Veigel et al. 1998), and the resulting speed, 0.2–20 µm/s, is close to the observed speeds. (Note that the above expression is similar to $v \sim d/t_s$, to within a factor of 2, the expression for the negligible drag case, but the implied molecular mechanism is different. $v \sim 2f_d/kt_s$ is also similar to v~gh, where g is the detachment rate and h is the maximum strain, derived by other workers (e.g., Huxley 1957: Cuda et al. 1997; Cooke et al. 1994, where "negatively strained" cross-bridges are the specific source of drag).

To explain the data presented according to such a model, t_s would necessarily decrease as $[Ca^{2+}]$ is increased. Physically, what can cause strong binding time to increase? One possibility is that Tm makes it harder for a bound myosin to dissociate at low $[Ca^{2+}]$, e.g., after Ca dissociates and Tm tries to swing back into place, it runs into the myosin head and makes it harder for it to detach. Electron diffraction studies (Vibert et al. 1997) show that myosin keeps Tm from its "off" position. On the other hand, Tm might push off the myosin head as it swings back into the "off" position, resulting in the opposite prediction. Clearly, more studies are needed to understand the dependence of speed on $[Ca^{2+}]$.

Note that in both cases above, it is implicitly assumed that the actin is a rigid rod, so that the whole filament's motion is correlated. The validity depends on the persistence length, l_p, of the regulated filament under the motility conditions. Measurements of l_p for actin filaments have ranged from as low as 3 µm (Orlova and Egelman 1993) to as long as 17 µm (Yanagida and Oosawa 1978; Yanagida et al. 1984; Gittes et al. 1993; Ott et al. 1993; Isambert et al. 1995). The experiment which yielded the shorter l_p relied on measurements of the contour length of a filament and end-to-end distance from electron micrographs (Orlova and Egelman 1993). This method can be complicated by the adhesive interaction between the filament and microscope grid and shrinkage effects, so that, at best, l_p's obtained from this method would be minimum values.

The measurements that yielded the longer l_p were based on Fourier analysis of the thermal fluctuations of filaments in the appropriate buffer. The l_p's obtained with this technique depend on the solutions used: F-actin's is ~17–20 µm; in the presence of HMM, this becomes ~12 µm, and in the presence of ATP, Ca^{2+}, Tn's and Tm's, the measured persistence length is ~9–12 µm (Yanagida and Oosawa 1978; Yanagida et al. 1984; Isambert et al. 1995). Since this technique is performed in environments closer to those in motility assays, the l_p's obtained would be more representative of the intrinsic l_p.

The lower end of our filament length groups are comfortably in the proper l_p range. Although the longest filament group may exceed l_p, comparisons with shorter filaments show no significant difference in behavior.

The above treatment is probably oversimplified and a combined analysis involving the physical and chemical interactions involved in the actin-Tm-myosin interface would be required to explain the data.

The observation that in vitro motility gliding speed is graded by [Ca^{2+}] is at least in qualitative agreement with recent skinned fiber data on unloaded shortening velocity (e.g., Moss 1992; Martyn et al. 1994; Metzger 1996). Using the slack method (e.g., Martyn et al. 1994), it has been shown that unloaded shortening speed, derived from the slope of the change in length vs slack time, is nonlinear and shows apparently biphasic behavior. Below a certain time the data show a slope ("rapid phase"), and above this region, the slope decreases ("slow phase"). Both phases show a dependence on [Ca^{2+}]. The origin for the biphasic behavior is not clear. It is not clear yet if in vitro motility corresponds to the early or late phase. In any case, it has been conjectured that the unloaded shortening speed in skinned fibers decreases with [Ca^{2+}] because of "internal" loads which must be overcome for the fiber to shorten or the filaments to slide. These loads could be passive (Moss1992) as produced by sarcomeric constraints or other proteins in the fiber, but these would be eliminated in the in vitro motility assay. These loads could be active such as friction caused by decreased detachment rates, when a cross-bridge is negatively strained. Further experiments are required in order to understand better the source of the [Ca^{2+}] dependence of the unloaded shortening speed.

5 Conclusion

The in vitro motility assay was utilized in an attempt to understand how calcium regulates muscle contraction by measuring the thin filament length dependence of the speed vs pCa. No length dependence in speed of moving portions was observed for 1- to >12-μm-long filaments, but the shortest regulated filaments were more likely to stop moving. The results were analyzed in the context of whether drag is significant or not. In both analyses, the data suggest that [Ca^{2+}] is possibly controlling the strong binding time, in addition to controlling the number of motors that bind. More work is needed to further clarify these results.

Acknowledgements. The authors acknowledge advice and assistance from Drs. M. Regnier, J. Howard, D. Martyn, E. Homsher; biochemical support from Drs. Y. Chen, C. Luo, C.-K. Wang; data analysis support from M. Mathiason, M. Woare, S. Velamoor; solution support from R. Mondares; and technical assistance from D. Anderson. This research is supported by NIH grant HL52558.

References

Bagshaw CR (1993) Muscle contraction. 2nd ed. Chapman and Hall

Bing W, Fraser IDC, Marston SB (1997a) Troponin I and troponin T interact with troponin C to produce different Ca^{2+}-dependent effects on actin-tropomyosin filament motility. Biochem J 327:335–340

Bing W, Redwood CS, Purcell IF, Esposito G, Watkins H, Marston SB (1997b) Effects of two hypertrophic cardiomyopathy mutations in a-tropomyosin, Asp175Asn and Glu180Gly, on Ca^{2+} regulation of thin filament motility. Biochem Biophys Res Commun 236:760–764

Bobkov AA, Bobkova EA, Homsher E, Reisler E (1997) Activation of regulated actin by SH1-modified myosin subfragment 1. Biochemistry 36:7733–7738

Brandt, PW, Diamond MS, Schachat FH (1984) The thin filament of vertebrate skeletal muscle co-operatively activates as a unit. J Mol Biol 180:379–384

Brenner B, Eisenberg E (1986) Rate of force generation in muscle: correlation with actomyosin ATPase activity in solution. Proc Natl Acad Sci USA 83:3542–3546

Cooke R, White H, Pate E (1994) A model of the release of myosin heads from actin in rapidly contracting muscle fibers Biophys J 66:778–788

Cuda G, Pate E, Cooke R, Sellers JR (1997) In vitro actin filament sliding velocities produced by mixtures of different types of myosin. Biophys J 72:1767–1779

ExpertVision (1990) Motion Analysis Corporation. 3650 N. Laughlin Road, Santa Rosa, CA 95403, USA

Farah CS, Reinach FC (1995) The troponin complex and regulation of muscle contraction. FASEB J 9:755–767

Finer JT, Simmons RM, Spudich JA (1994) Single myosin molecule mechanics: picoNewton forces and nanometer steps. Nature 368:113–119

Ford LE, Huxley AF, Simmons RM (1985) Tension transients during steady shortening of frog muscle fibres. J Physiol 361:131–150

Fraser IDC, Marston SB (1995) In vitro motility analysis of actin-tropomyosin regulation by troponin and calcium: the thin filament is switched as a single cooperative unit. J Biol Chem 270:7836–7841

Gerson JH, Bobkova E, Homsher E, Reisler E (1997) Role of residue 311 in actin-tropomyosin interaction: in vitro motility study using yeast actin mutant E311 A/R312 A. Biophys J 72:A57 (Abstr)

Gittes F, Mickey B, Nettleton J, Howard J (1993) Flexural rigidity of microtubules and actin filaments measured from thermal fluctuations in shape. J Cell Biol 120:923–934

Gordon AM, Godt RE, Donaldson SKB, Harris CE (1973) Tension in skinned frog muscle fibers in solutions of varying ionic strength and neutral salt composition. J Gen Physiol 62:550–574

Gordon AM, LaMadrid MA, Chen Y, Luo Z, Chase PB (1997) Calcium regulation of skeletal muscle thin filament motility in vitro. Biophys J 72:1295–1307

Gordon AM, Chen Y, Liang B, LaMadrid MA, Luo Z, Chase PB (1998) Skeletal muscle regulatory proteins enhance F-actin in vitro motility. Adv Exp Med Biol 453:187–196

Gordon AM, Homsher E, Regnier M (2000) Regulation of contraction in striated muscle. Physiol Rev 80:853–924

Greene LE, Williams DL, Eisenberg E (1987) Regulation of actomyosin ATPase activity by troponin-tropomyosin: effect of the binding of the myosin subfragment 1 (S-1)ATP complex. Proc Natl Acad Sci USA 84:3102–3106

Griffiths PJ, Ashley CC, Bagni MA, Maeda Y, Cecchi G (1993) Cross-bridge attachment and stiffness during isotonic shortening of intact single muscle fibers. Biophys J 64:1150–1160

Gulati J (1983) Magnesium ion-dependent contraction of skinned frog muscle fibers in calcium-free solution. Biophys J 44:113–121

Harada Y, Sakurada K, Aoki T, Thomas D, Yanagida T (1990) Mechanochemical coupling in actomyosin energy transduction studied by in vitro movement assay. J Mol Biol 216:49–68

Harris DE, Warshaw D (1993) Smooth and skeletal muscle myosin both exhibit low duty cycles at zero load in vitro. J Biol Chem 268:14764–14768

Homsher E, Wang F, Sellers JR (1992) Factors affecting movement of F-actin filaments propelled by skeletal muscle heavy meromyosin. Am J Physiol 262:C714–C723

Homsher E, Kim B, Bobkova A, Tobacman LS (1996) Calcium regulation of thin filament movement in an in vitro motility assay. Biophys J 70:1881–1892

Honda H, Asakura S (1989) Calcium-triggered movement of regulated actin in vitro A fluorescence microscopy study. J Mol Biol 205:677–683

Honda H, Tamura T, Hatori K, Matsuno K (1995) Decorating actin filaments with troponin T-I complexes and acceleration of their sliding movement on myosin molecules. Biochim Biophys Acta 1251:43–47

Honda H, Kitano Y, Hatori K, Matsuno K (1996) Dual role of tropomyosin on chemically modified actin filaments from skeletal muscle. FEBS Lett 383:55–58

Honda H, Tagami N, Hatori K, Matsuno K (1997) Regulated crosslinked actin filaments and the decoupling between their ATPase activity and sliding motility. J Biochem Tokyo 121:47–49

Howard J (1997) Molecular motors: structural adaptations to cellular functions. Nature 389: 561–567

Huxley AF (1957) Muscle structure and theories of contraction. Prog Biophys Chem 7:255–318

Isambert H, Venier P, Maggs A, Fattoum A, Kassab R, Pantaloni D, Carlier MF (1995) Flexibility of actin filaments derived from thermal fluctuations. J Biol Chem 270:11437–11444

Julian FJ, Morgan DL (1981) Variation of muscle stiffness with tension during tension transients and constant velocity shortening in the frog. J Physiol 319:193–203

Kron SJ, Toyoshima YY, Uyeda TQP, Spudich JA (1991) Assays for actin sliding movement over myosin-coated surfaces. Methods Enzymol 196:399–416

Landis CA, Bobkova A, Homsher E, Tobacman LS (1997) The active state of the thin filament is destabilized by an internal deletion in tropomyosin. J Biol Chem 272:14051–14056

Lee D, Bobkova A, Tobacman LS, Homsher E (1997) The use of pPDM-HMM and microneedles to measure pN forces exerted by myosin heads on single regulated thin filaments. Biophys J 72:A57

Lee DM, Tobacman LS, Homsher E (1998) Calcium dependence of isometric force measured in single thin filaments using glass microneedles. Biophys J 74:A347

Lehrer SS, Morris EP (1982) Dual effects of tropomyosin and troponin-tropomyosin on actomyosin subfragment 1 ATPase. J Biol Chem 257:8073–808

Leibler S, Huse DA (1993) Porters versus rowers: a unified stochastic model of motor proteins. J Cell Biol 121:1357–1680

Lin D, Bobkova A, Homsher E, Tobacman LS (1996) Altered cardiac troponin T in vitro function in the presence of a mutation implicated in familial hypertrophic cardiomyopathy. J Clin Invest 97:2842–2848

Martyn M, Chase PB, Hannon J, Huntsman L, Kushmerick M, Gordon AM (1994) Unloaded shortening of skinned muscle fibers from rabbit activated with and without Ca^{2+}. Biophys J 67: 1984–1993

Mehta AD, Finer JT, Spudich JA (1997) Detection of single molecule interactions using correlated thermal diffusion. Proc Natl Acad Sci USA 94:7927–7931

Metzger JM (1996) Effects of phosphate and ADP on shortening velocity during maximal and submaximal calcium activation of the thin filament in skeletal muscle fibers. Biophys J 70: 409–417

Molloy JE, Burnes JE, Sparrow JC, Tregear RT, Kendrick-Jones J, White DCS (1995a) Single molecule mechanics of heavy meromyosin and S1 interacting with rabbit or *Drosophila* actins using optical tweezers. Biophys J 68:298s–305s

Molloy JE, Burns JE, Kendrick-Jones J, Tregear RT, White DCS (1995b) Force and movement produced by a single myosin head. Nature 378:209–212

Moss RL (1992) Ca^{2+} regulation of mechanical properties of striated muscle. Mechanistic studies using extraction and replacement of regulatory proteins. Circ Res 70:865–884

Orlova A, Egelman EH (1993) A conformational change in the actin subunit can change the flexibility of the actin filament. J Mol Biol 232:334–341

Ott A, Magnasco M, Simon A, Libchaber A (1993) Measurement of the persistence length of polymerized actin using fluorescence microscopy. Phys Rev E 48:R1642–1645

Regnier M, Martyn DA, Chase PB (1998) Calcium regulation of tension redevelopment kinetics with 2-deoxy-ATP or low [ATP] in rabbit skeletal muscle. Biophys J 74:2005–2015

Sase I, Miyata H, Ishiwata S, Kinosita, K Jr (1997) Axial rotation of sliding actin filaments revealed by single fluorophore imaging. Proc Natl Acad Sci USA 94:5646–5650

Sata M, Sugiura S, Yamashita H, Aoyagi T, Momomura S, Serizawa T (1995a) Pimobendan directly sensitizes reconstituted thin filament to slide on cardiac myosin. Eur J Pharmacol 290:55–59

Sata M, Sugiura S, Yamashita H, Fujita H, Momomura S, Serizawa T (1995b) MCI-154 increases Ca^{2+} sensitivity of reconstituted thin filament. A study using a novel in vitro motility assay technique. Circ Res 76:626–633

Sata M, Yamashita H, Sugiura S, Fujita H, Momomura S, Serizawa T (1995c) A new in vitro motility assay technique to evaluate calcium sensitivity of the cardiac contractile proteins. Pfluegers Arch 429:443–445

M. Schoenberg (1998) Characterization of the myosin adenosine triphosphate (M.ATP) crossbridge in rabbit and frog skeletal muscle fibers. Biophys J 54:135–148

Sellers JR, Cuda G, Wang F, Homsher E (1993) Myosin specific adaptations of the motility assay. Methods Cell Biol 39:23–49

Squire JM, Morris EP (1998) A new look at thin filament regulation in vertebrate skeletal muscle. FASEB J 12(10):761–771

Szent-Györgyi AG (1975) Calcium regulation of muscle contraction. Biophys J 15:707–723

Tawada K, Sekimoto K (1991) A physical model of ATP-induced actin-myosin movement in vitro. Biophys J 59:343–356

Tobacman LS (1996) Thin filament-mediated regulation of cardiac contraction. Annu Rev Physiol 58:447–481

Toyoshima YY, Kron SJ, McNally EM, Niebling KR, Toyoshima C, Spudich JA (1987) Myosin subfragment-1 is sufficient to move actin filaments in vitro. Nature 328:536–539

Uyeda TQP, Kron SJ, Spudich JA (1990) Myosin step size estimation from slow sliding movement of actin over low densities of heavy meromyosin. J Mol Biol 214:699–710

Veigel C, Bartoo ML, White DC, Sparrow JC, Molloy JE (1998) The stiffness of rabbit skeletal actomyosin cross-bridges determined with an optical tweezers transducer. Biophys J 75(3): 1424–1438

Vibert P, Craig R, Lehman W (1997) Steric model for activation of muscle thin filaments. J Mol Biol 266:8–14

Williams DL, Greene LE, Eisenberg E (1988) Cooperative turning on of myosin subfragment 1 adenosinetriphosphatase activity by the troponin-tropomysoin-actin complex. Biochemistry 27:6987–6993

Xie X, Harrison DH, Schlichting I, Sweet RM, Kalabokis VN, Szent-Györgyi AG, Cohen C (1994) Structure of the regulatory domain of scallop myosin at 2.8-Å resolution. Nature 368:306–312

Yanagida T, Oosawa F (1978) Polarized fluorescence from e-ADP incorporated into F-actin in a myosin-free single fiber: conformation of F-actin and changes induced in it by heavy meromyosin. J Mol Biol 126:507–524

Yanagida T, Nakase M, Hishiyama K, Oosawa F (1984) Direct observation of motion of single F-actin filaments in the presence of myosin. Nature 307:58–60

The Ultrastructural Basis of Actin Filament Regulation

Roger Craig[1] and William Lehman[2]

1 Introduction

Actin filaments are found in virtually all cells, performing diverse functions throughout the animal and plant kingdoms. They are generally dynamic structures that may disassemble or reassemble as needed. Actin accomplishes these astonishing feats by binding to specific actin-binding proteins, which modify its structure, enabling it to perform the required function. These actin-binding proteins include components that cross-link filaments (forming bundles or networks), sever filaments (causing depolymerization), sequester actin monomers (preventing polymerization), cap filaments (inhibiting depolymerization), and those that act as motor proteins to generate motility.

In striated muscle, actin filaments are permanent structures that are prevented from depolymerizing by capping proteins at both Z-line and M-line ends. They perform two crucial functions in contraction: (1) They undergo cyclic interaction with myosin crossbridges, generating sliding of the actin filaments past the myosin filaments, which leads to shortening and force. (2) In most striated muscles, they also *regulate* contraction, by switching crossbridge interaction with actin ON or OFF, in response to changes in cytosolic free Ca^{2+} concentration. Regulation in most striated muscles is dependent on association of actin filaments with the actin-binding protein, tropomyosin (Tm), which in turn is associated with the Ca^{2+}-binding protein complex, troponin (Tn). In smooth muscles, tropomyosin is associated with other actin-binding proteins, which may be involved in modulating actin-myosin interaction.

In this chapter we review the contributions of electron microscopy (EM) and image processing to our understanding of the molecular mechanism of actin filament regulation in striated and smooth muscles. We correlate the results with structural data obtained from X-ray diffraction, and we relate interpretations of structural data to models based on biochemical and other approaches. For an earlier, comprehensive review on ultrastructural studies of thin filaments, see O'Brien and Dickens (1983).

[1] Department of Cell Biology, University of Massachusetts Medical School, 55 Lake Avenue North, Worcester, Massachusetts 01655, USA

[2] Department of Physiology and Biophysics, Boston University School of Medicine, 715 Albany Street, Boston, Massachusetts 02118, USA

Results and Problems in Cell Differentiation, Vol. 36
D.D. Thomas, C.G. dos Remedios (Eds.): Molecular Interactions of Actin

2 Regulation in Striated Muscle

2.1 Background

The G-actin monomer (42 kDa) polymerizes, both in vitro and in vivo, to form actin *filaments* with a double helical organization of monomer subunits (Fig. 1a). This structure was first clearly demonstrated by electron microscopy of negatively stained filaments polymerized in vitro from purified actin and of native filaments isolated directly from muscle (Hanson and Lowy 1963; Huxley 1963). These observations distinguished between two similar structures that had been proposed on the basis of X-ray diffraction of muscle (Selby and Bear 1956; see O'Brien and Dickens 1983). The filaments appeared as two strings of beads (roughly spherical actin monomers) twisted around each other to form two long-pitch helices with crossovers every 36–38 nm (Fig. 1b). The subunit repeats along each helix were approximately 5.5 nm, and the two strings were staggered by approximately 1/2 repeat (2.75 nm). The two helices were shown by unidirectional shadowing to twist around each other in a right-handed sense (Depue and Rice 1965). This structure is defined more formally by a left-handed "genetic" helix (a single helix connecting every subunit) with pitch 5.9 nm (see O'Brien and Dickens 1983). This basic structural framework formed the foundation for all subsequent studies of actin filaments and is central to understanding the structural basis of actin filament regulation.

As mentioned above, two other proteins are associated in abundance with the actin filaments of striated muscle, the actin-binding protein tropomyosin, and the Ca^{2+}-binding protein complex, troponin (reviewed by Weber and Murray 1973; El-Saleh et al. 1986; Tobacman 1996). The essential role of these proteins in regulating actin-myosin interaction was demonstrated in Ebashi's classic biochemical experiments (reviewed by Ebashi et al. 1969), showing that "native tropomyosin", a complex of tropomyosin and troponin, was required for Ca^{2+}-regulation of actin-activated myosin ATPase activity (actin-myosin ATPase is the biochemical analog of the actin-myosin cross-bridge cycle that drives contraction). Pure actin activates myosin ATPase independently of Ca^{2+}. When troponin and tropomyosin are present (in native or reconstituted thin filaments) this activation is inhibited at low Ca^{2+}, and the inhibition is reversed at high Ca^{2+} (Weber and Murray 1973).

Tropomyosin is a coiled-coil protein formed from two parallel 35-kDa α-helical subunits (reviewed by Cohen 1975; O'Brien and Dickens 1983). This elongated molecule (42 nm long, 2 nm wide) extends along the long-pitch actin helices, an arrangement that was first proposed on the basis of the elongated shape of tropomyosin (Hanson and Lowy 1963) and the newly deduced structure of the actin helix, although tropomyosin was not itself visible. This organization was supported by subsequent X-ray diffraction and EM studies (e.g. Hanson et al. 1973), and in well-stained specimens it is now possible to *directly* visualize tropomyosin following the actin helices (Fig. 1c). Tropomyosin

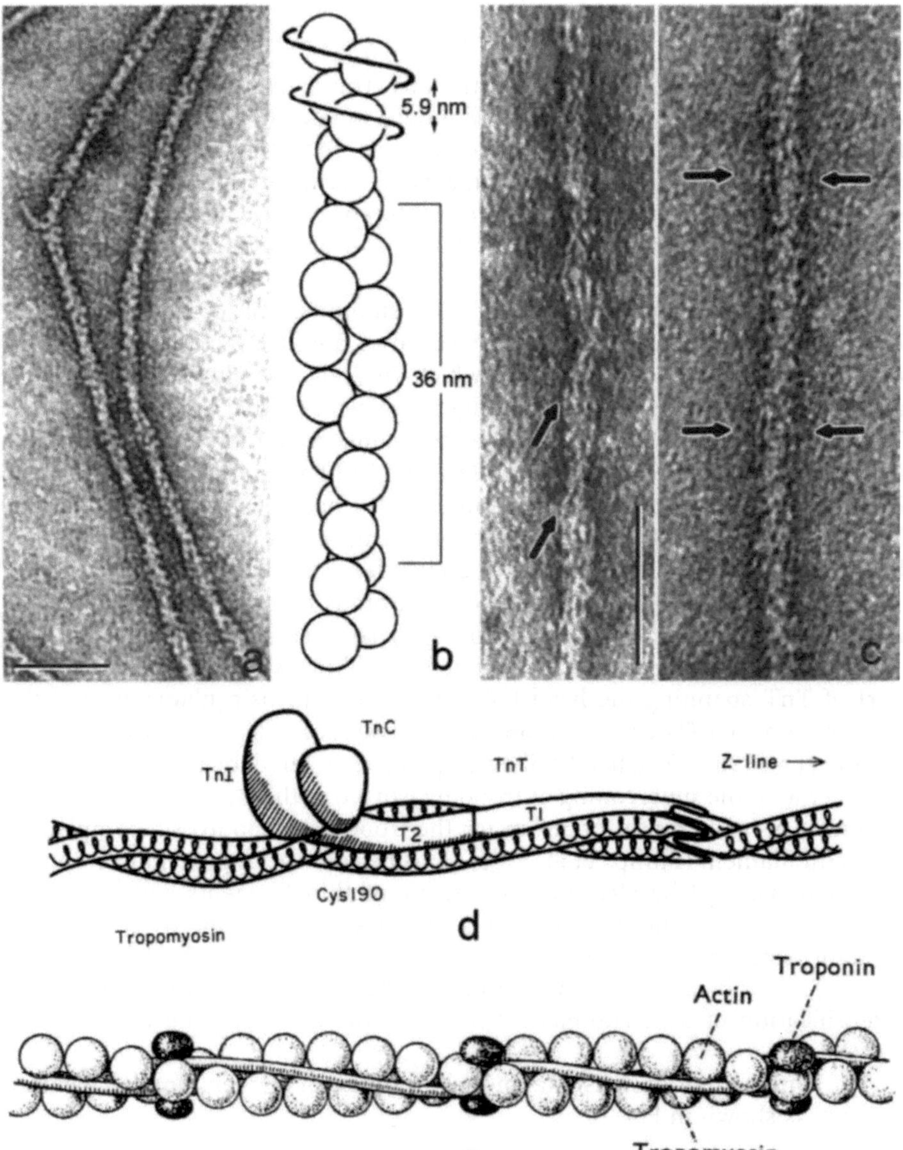

Fig. 1a–e. Early concepts of thin filament structure. **a** Filaments of pure F-actin negatively stained with uranyl acetate (modified from Vibert and Craig 1982, with permission). **b** Original model of F-actin based on negative stain images (Hanson and Lowy 1963), with actin subunits portrayed as spheres. Main features of helical symmetry include the 36-nm "crossover" repeat of the two long-pitch helices, and the left-handed genetic helix of pitch 5.9 nm. **c** Negatively stained thin filaments, selected to illustrate direct visualization of tropomyosin strands (*arrows*) following the long-pitch actin helices. *Left* Cardiac thin filament (taken from Lehman et al. 1995, with permission); *right* smooth muscle thin filament (taken from Moody et al. 1990, with permission). **d** Model of troponin and its association with tropomyosin, based on rotary shadowing (taken from Flicker et al. 1982, with permission). **e** Early model of thin filament showing proposed arrangement of tropomyosin and troponin on F-actin (taken from Ebashi et al. 1969, with permission). *Scale bars* = 50 nm

molecules polymerize to form a continuous strand, one associated with each actin helix, with a head-tail overlap of about 1 nm occurring between the non-helical tips of the molecules. One tropomyosin molecule extends over seven actin monomers, making similar contacts on each, corresponding to exactly seven half turns of the coiled-coil and seven repeating groups of charged and non-polar residues along its length (Phillips et al. 1986; cf. McLachlan and Stewart 1976).

The troponin complex consists of three subunits: TnI (23 kDa), which can *I*nhibit actomyosin ATPase in the presence or absence of tropomyosin, and independently of Ca^{2+}; TnT (37 kDa), which binds the complex to *T*ropomyosin; and TnC (18 kDa), the **C**alcium-binding component (Weber and Murray 1973). The structure of TnC has been solved to atomic resolution by X-ray crystallography, although the structures of TnT and TnI have not. By rotary shadowing electron microscopy, the troponin complex has been found to be 26-nm long (more than half the length of tropomyosin), with a 2-nm-wide tail 16-nm long, comprising mainly TnT, and a globular head consisting of TnI and TnC (Flicker et al. 1982; Fig. 1d). Troponin binds to tropomyosin with a 1:1 molar ratio. EM and X-ray crystallography suggest that the elongated TnT tail extends along the tropomyosin coiled-coil (ibid), with the amino-terminal part of TnT spanning the head-tail joint of tropomyosin filaments, and the head region (TnI/TnC) 20-nm away, near residues 150–180 of tropomyosin (White et al. 1987). The head-tail overlap between neighboring tropomyosins, together with the supercoiling of tropomyosin along the long-pitch actin helix, generates an axial repeat of 38 nm for the troponin-tropomyosin complex on the actin filament (Philips et al. 1986). This periodic arrangement of troponin was demonstrated by electron microscopy of thin filaments and myofibrils labeled with antibodies to troponin (Ohtsuki et al. 1967) and by comparison of actin paracrystals with actin-tropomyosin-troponin paracrystals (Hanson 1973). X-ray diffraction of muscle fibers labeled with anti-troponin showed intensification of a meridional reflection at the same spacing (Rome et al. 1973).

The simple three-dimensional model for the actin filament implied by these findings is shown in Fig. 1e.

2.2 The "Steric Blocking" Model of Regulation

The current working structural model for regulation of actin filament interaction with myosin is the "steric blocking" mechanism. In its simplest form, this model involves the physical blocking of the myosin binding site on actin by tropomyosin under low Ca^{2+} conditions ($<10^{-8}$ M). Because myosin heads are prevented from binding to actin filaments, the cross-bridge cycle is inhibited, and muscle is therefore relaxed. At high Ca^{2+} levels ($>10^{-6}$ M), the binding of Ca^{2+} by troponin causes a uniform azimuthal shift in the position of the entire tropomyosin molecule towards what was originally considered to be the "groove" between the two actin helices (Fig. 2a) so that the myosin binding sites on the seven actin subunits to which each tropomyosin binds are uncovered, allowing the cyclic interaction of myosin crossbridges with actin, and thus filament sliding, to occur. This model accounts for the co-operative switching ON of thin filaments, in which Ca^{2+} binding by troponin switches on seven actins (Bremel and Weber 1972), a single troponin molecule regulating seven actin subunits through its association with the extended tropomyosin.

This model was originally proposed on the basis of the positions of tropomyosin and myosin heads that had been inferred from 3-D reconstructions of thin filaments (Moore et al. 1970; Spudich et al. 1972), and their relationship to changes in the X-ray diffraction pattern of contracting muscle compared with relaxed muscle (Haselgrove 1973; Huxley 1973; Parry and Squire 1973). During contraction, the intensity of the second (19 nm) layer line coming from the actin filaments was found to increase. Assuming that tropomyosin followed the long-pitch actin helices, this change could be simply accounted for by a movement of tropomyosin from a position on one side of the actin "groove" to one closer to the center (Fig. 2a).

Subsequent studies supported this model of Ca^{2+}-sensitive tropomyosin movement. X-ray diffraction of oriented gels of reconstituted actin-tropomyosin-troponin filaments showed Ca^{2+}-sensitive changes in the actin layer lines similar to those occurring in intact muscle (Gillis and O'Brien 1975; Popp and Maéda 1993), and image processing and 3-D reconstruction of negatively stained paracrystals of thin filaments gave similar results (Gillis and O'Brien 1975; Wakabayashi et al. 1975). Later, millisecond time-resolved studies of contracting muscle showed that the layer line changes occurred before muscle tension rose, consistent with the proposal that tropomyosin movement was required before crossbridges could generate tension (Kress et al. 1986).

2.3 Early Controversies Concerning the Steric Model

Appealing though the steric blocking model was, fundamental uncertainties remained (see Cohen and Vibert 1987 for review). One criticism came from transient kinetic studies of actin filaments in solution, which showed that the rate of binding of myosin heads to regulated actin filaments was independent of Ca^{2+} concentration, even though actin-activated myosin ATPase *was* regulated (see Chalovich 1992). A modification to the steric blocking model was

---

The term "groove" was used in early studies of actin filaments to describe the dip that occurred between the two long-pitch helices, at the time that the actin monomer was considered to be a simple sphere (Figs. 1b,e, 2a). With our current knowledge of the two-domain shape of actin, there is no longer a pronounced groove between the helices, and this term is therefore no longer appropriate (see Figs. 2b–e).

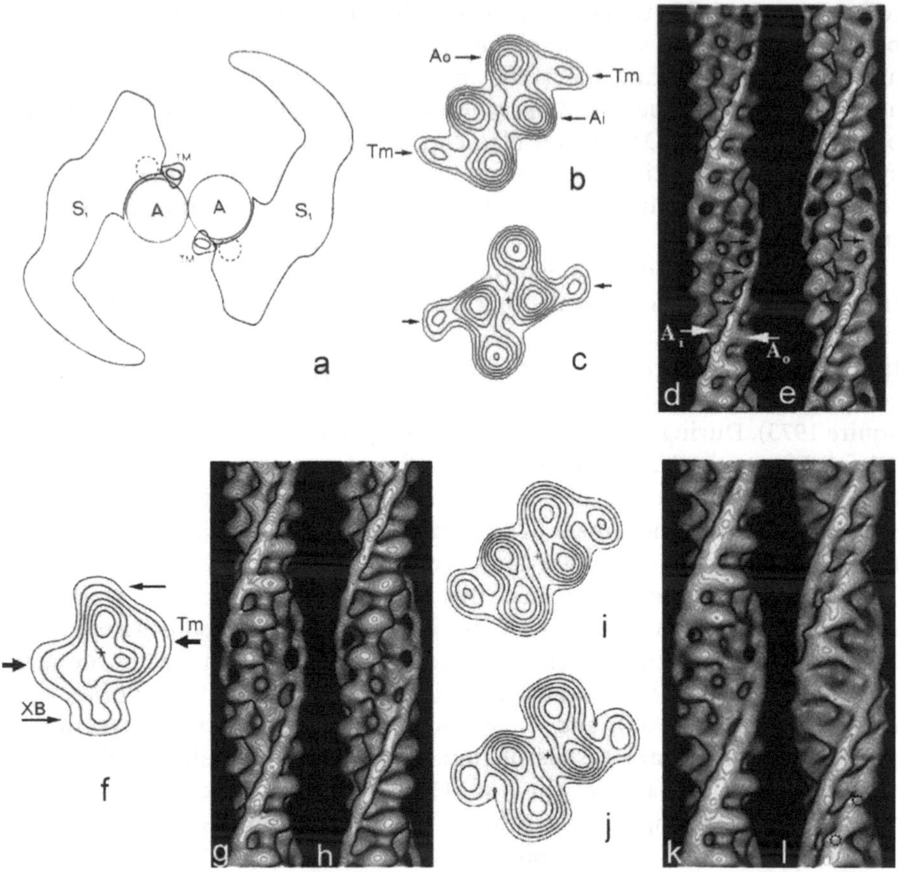

Fig. 2. **a** Early model of steric blocking, based on X-ray diffraction and early reconstruction data, showing thin filament in transverse view. *A* Actin, S_1 myosin head; the proposed low and high Ca^{2+} positions of tropomyosin (*Tm*) are shown by *dotted* and *solid lines* respectively (taken from Huxley 1973, with permission). **b–l** 3-D reconstructions of thin filaments. **b** and **c** Helical projections (projections down the two long-pitch actin strands) of *Limulus* thin filaments showing tropomyosin (*arrows*) attached to **b** the outer domain of actin (A_o) at low Ca^{2+} and **c** the inner domain (A_i) at high Ca^{2+}. **d** and **e** Surface views of *Limulus* thin filament reconstructions showing tropomyosin strand on the inner edge of the outer domain of actin at low Ca^{2+} (**d**) and on the inner domain at high Ca^{2+} (**e**). **f** Transverse section through reconstruction of thin filament interacting with myosin filament during filament sliding, showing tropomyosin (*Tm, thick arrows*) on inner actin domain, and cross-bridge density *(XB, thin arrows)* on outer actin domain; **g** surface view of actin filament containing smooth muscle tropomyosin, seen as a strand connecting to the outer domain of actin; **h** same as (**g**) but with caldesmon also present – tropomyosin has now been displaced to the inner domain; **i, j** helical projections of filaments in **g, h** respectively showing connection of tropomyosin to outer actin domain in the absence of caldesmon (**i**) and the inner domain in its presence (**j**); **k, l** surface views of actin-smooth muscle tropomyosin filaments in the absence (**k**) and presence (**l**) of calponin. Tropomyosin is on the outer domain in (**k**) and on the inner domain in (**l**); additional density of calponin in (**l**) is marked with *asterisks*. **b–e** modified from Vibert et al., (1997); **g–j** modified from Hodgkinson et al. (1997a); **k–l** taken from Hodgkinson et al. (1997b), all with permission

therefore proposed in which tropomyosin did not block the *initial binding* of myosin heads at low Ca^{2+}, but did block the subsequent motion of the heads required for force generation.

There were also a number of structural uncertainties and disagreements. X-ray fiber diffraction provides only indirect data on structure and had to be interpreted by model-building (i.e., by comparison of experimental X-ray data with computed diffraction patterns of different models). A unique interpretation of structure was therefore not possible. Electron microscopy does provide direct images, but these too have potential problems. Negative staining can produce staining or drying artifacts, and the electron beam can cause stain migration giving a false outline of protein surfaces. Most early EM observations were based on thin filament paracrystals (two- or three-dimensional arrays of parallel filaments), in which interfilament bonds may have altered filament structure, and protein boundaries could not be reliably established. Many of the EM reconstructions of the 1970s and early 1980s contained incorrectly positioned densities owing to an error in the reconstruction program that was used (discussed by Cohen and Vibert 1987). Finally, all reconstructions reveal only mass densities, and in multi-protein complexes such as thin filaments, assignments to specific proteins are largely subjective; this problem was compounded by the limited resolution of the early reconstructions such that fine but crucial features of structure may have been missing. An example of this was the controversy concerning the assignment of densities to myosin heads, actin and tropomyosin in early reconstructions, and hence uncertainty as to whether tropomyosin was on the "correct" side of the actin monomer to directly block myosin head binding (Taylor and Amos 1981; Seymour and O'Brien 1980, 1985).

2.4 Recent Advances in the Ultrastructural Basis of Regulation

Recent improvements in experimental technique, which avoid almost all of the above problems, have resolved most of these uncertainties and have greatly strengthened support for the steric blocking model. Cryo-electron microscopy of thin filaments decorated with myosin heads (subfragment-1) produced 3-D reconstructions showing that tropomyosin lies on the same side of the actin as the myosin binding site (Milligan and Flicker 1987; Milligan et al. 1990). Actin was now resolved into two domains (in agreement with the X-ray crystal structure of actin; Kabsch et al. 1990), with the long axis of the monomer roughly perpendicular to the filament axis. S1 was seen to bind to actin tangentially (as originally proposed by Moore et al. 1970), covering the front concave face of the actin monomer, mostly on the outer domain). Tropomyosin in these decorated filaments was seen to be bound on the inner domain of actin. However, data on undecorated filaments were not presented, so the high and low Ca^{2+} positions of tropomyosin in the absence of bound heads predicted by X-ray diffraction could not be confirmed.

A critical further advance was made using negatively stained native thin filaments isolated directly from *Limulus* (horseshoe crab) striated muscle, which were observed with low electron doses to minimize beam damage (Lehman et al. 1994). In these images, tropomyosin was revealed *directly* (without image processing), and was seen to run along the long-pitch actin helices (cf. Fig. 1c). Reconstructions were computed from single filaments stained under high and low Ca^{2+} conditions (thus avoiding the uncertainties arising from earlier analyses of paracrystalline arrays), and the data from multiple filaments were averaged so that true densities could be distinguished from noise. These reconstructions revealed for the first time the position of tropomyosin in both the OFF and the ON positions (Fig. 2b–e). The two-domain shape and monomer-monomer connectivity of actin subunits were indistinguishable from those in the frozen-hydrated map described above, demonstrating that artifacts of negative staining, if present, were at most minor. At low Ca^{2+} (the OFF position) tropomyosin was seen to lie on the inner edge of the outer domain of actin (Fig. 2b,d), coinciding approximately with the site occupied by myosin heads when they bind strongly to actin in the absence of ATP (Rayment et al. 1993). Following treatment with high Ca^{2+}, tropomyosin moved azimuthally by 25° (2 nm) to contact actin on the inner domain (the ON position, Fig. 2c,e). These ON and OFF positions of tropomyosin, and their relation to the site of strong myosin binding on actin, lent strong support to a steric mechanism in which tropomyosin in relaxed muscle could physically block the transition from an initially weak to a strong myosin cross-bridge attachment to actin (without necessarily blocking weak initial attachment) thus inhibiting actomyosin ATPase and cross-bridge cycling (Chalovich 1992).

There remained some uncertainty as to whether the results from the horseshoe crab (a marine invertebrate with a larger than usual troponin molecule) would apply more generally. Experiments on frog striated muscle showed that they did. Reconstructions from frog skeletal muscle thin filaments at low Ca^{2+} showed an identical position for tropomyosin, on the inner edge of the outer domain of actin (Lehman et al. 1995). Reconstructions computed for frog cardiac thin filaments also showed the same tropomyosin appearance and position. The similar appearances of the tropomyosin strands in skeletal and cardiac filaments demonstrated, in addition, that nebulin, a narrow, highly elongated protein running the entire length of skeletal muscle thin filaments (Trinick 1994), but confined to the Z-line end of cardiac filaments (Moncman and Wang 1995), did not contribute significantly to the strand interpreted as tropomyosin. Moreover, negative stain reconstructions of filaments reconstituted from purified actin, tropomyosin and troponin (without nebulin) have also recently been carried out, and appear identical to those of filaments isolated directly from muscle, supporting this interpretation (Xu et al. 1999).

2.5 Towards an Atomic Level Model for Regulation

The 3-D reconstructions computed from these negative stain and cryo-images are of sufficient quality and resolution that it is possible to computationally "fit" them to the atomic model of the actin filament determined by X-ray diffraction of actin crystals (Kabsch et al. 1990) and of oriented actin filament gels (Holmes et al. 1990; Lorenz et al. 1993). This powerful procedure effectively bridges the gap between the resolutions obtainable by X-ray crystallography (0.2 nm) and electron microscopy of filamentous structures (2 nm), producing a near-atomic level model of actin filament structure. Because filaments cannot be crystallized, this is the only direct procedure for generating near-atomic level models of filamentous structures. Fitting consists, in essence, of matching the shape and orientation of the actin monomer determined by X-ray crystallography to that of the subunits observed in EM reconstructions. Because actin is an asymmetric molecule with a central cleft that is visible by both techniques, the X-ray structure can be fitted uniquely to the filament subunits seen by EM. This process establishes within the reconstruction the positions of the four actin subdomains seen by X-ray crystallography, and the approximate positions of individual amino acids of actin.

Fitting the actin filament atomic model to 3-D reconstructions of the OFF state (Fig. 2d) reveals that tropomyosin contacts subdomain-1 of actin near its junction with subdomain-3, then bridges over, without touching subdomain-2, to neighboring subdomains 1 and 3 of the next monomer along the long-pitch helix (Fig. 3a; Lehman et al. 1995; Vibert et al. 1997). Several clusters of amino acid residues likely to be involved in strong, stereo-specific myosin binding in active muscle (Rayment et al. 1993; red in Fig. 3a) lie beneath tropomyosin and are presumably inaccessible to myosin in the OFF state. Sites of proposed weak, electrostatic interaction with myosin (green in Fig. 3a) are not covered. These results agree well with models of actin-tropomyosin structure based on X-ray diffraction data (Squire et al. 1993; Poole et al. 1994). In the ON state (high Ca^{2+}, Fig. 3b), the 2-nm movement of tropomyosin on to the inner domain of actin uncovers most of the strong binding clusters (now green), but still masks one (residues 332–336, red), located at the junction of subdomains 1 and 3. Thus Ca^{2+}-induced tropomyosin movement would facilitate cross-bridge binding, but would not *fully* switch on the thin filament (Vibert et al. 1997). Strong binding of myosin heads to actin (in the absence of ATP) causes an additional 1-nm shift of tropomyosin further on to the inner domain, exposing the remaining covered sites (Fig. 3c; Vibert et al. 1997).

A simple structural model emerges from these studies. At low Ca^{2+} levels, tropomyosin physically blocks the strong (specific) myosin binding site on each of the seven actin subunits with which it interacts. The putative nonspecific electrostatic sites of interaction on subdomain-1 are not affected. Thus, transient weak interactions of myosin crossbridges with actin that may occur in relaxed muscle would be possible, while transition to a strongly attached, force-producing state would be blocked (cf. Chalovich 1992). When

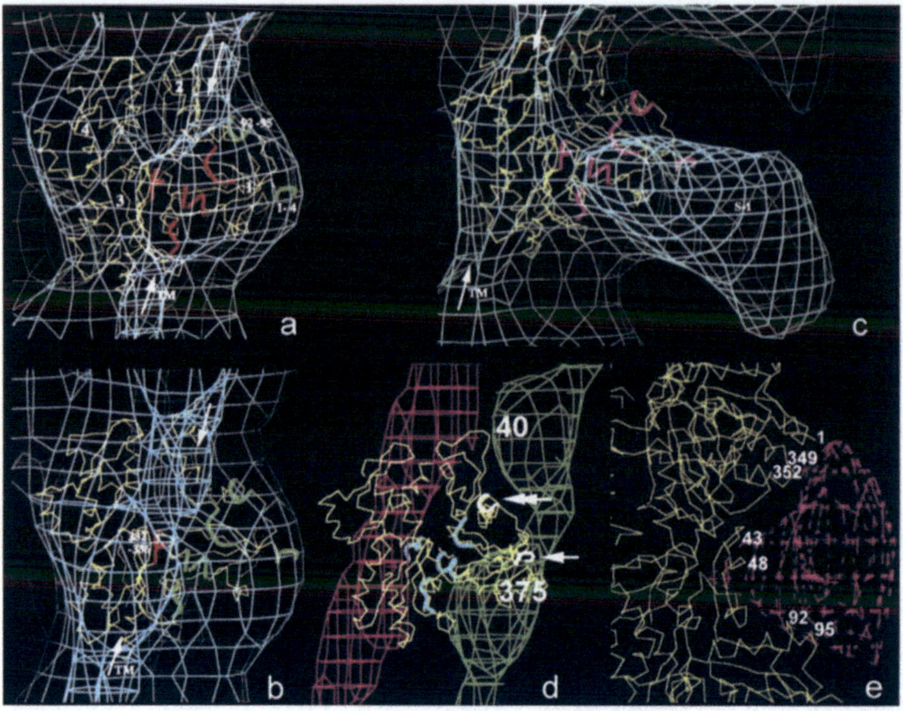

Fig. 3. Fitting of the atomic structure of actin (*yellow*, α-carbon chain) to thin filament reconstructions (*blue wire cage* in **a–c**). A single actin subunit is shown in each case. **a** Striated muscle thin filament at low Ca^{2+}. Tropomyosin (*TM, arrows*) contacts subdomain-1 of actin, lying over amino acid clusters involved in strong myosin binding (*red*), but leaving weak myosin binding sites exposed (*green*). **b** At high Ca^{2+}, tropomyosin moves on to subdomain-3, uncovering (*green*) all but one (*red*) of the strong myosin binding clusters. **c** When myosin heads (*S1*) are strongly bound (in the absence of ATP) tropomyosin moves further on to subdomain-3, uncovering the rest of the myosin binding site. **d–e** Fitting to smooth muscle thin filaments (wire cage has been omitted in these cases). **d** Single actin monomer (*yellow*) in actin-tropomyosin-caldesmon filament, showing location of tropomyosin (*red*) over the inner domain of actin and caldesmon (*green*) on the periphery of the outer domain. Residues involved in strong myosin binding (*blue*) are not covered. **e** Parts of two actin monomers (*yellow*) in actin-calponin filament, showing calponin (*pink*) attached to subdomain-1 of neighboring actins, and passing over the hollow formed by subdomain-2. **a–c** Taken from Vibert et al. (1997); **d** taken from Lehman et al. (1997); **e** taken from Hodgkinson et al. (1997b), all with permission

Ca^{2+} is released, tropomyosin movement would unblock most of the myosin binding site, facilitating binding of myosin heads, but not fully switching on contraction. Binding of some heads would cause further movement of tropomyosin, fully exposing the myosin binding site and facilitating the binding of additional heads. This would result in the fully activated state, with a high probability of myosin binding. This simple three-state structural model is qualitatively consistent with independently proposed models involving three regulatory states (Bremel and Weber 1972; Phillips et al. 1986; McKillop and Geeves 1993; Poole et al. 1995; al-Khayat et al. 1995; Lehrer and Geeves 1998). Reconstructions of filaments partially decorated with myosin heads (Vibert et al. 1997), show that as few as two heads per seven actins are sufficient to cause tropomyosin to move to the position in which it fully exposes the myosin binding site, and that tropomyosins up to three molecules away can also be moved. These structural observations are qualitatively consistent with the known cooperative behavior of tropomyosin.

The excellent fit that is obtained between the 3-D reconstructions and the crystallographic structure of actin, and the agreement between EM and X-ray models of thin filaments in different states, supports the validity of the reconstruction process and shows that negative staining preserves actin filament structure at near-atomic resolution. The most definitive results would come, however, from *cryo-EM* reconstructions of thin filaments under low and high Ca^{2+} conditions, where potential staining and drying artifacts would be completely eliminated. Such studies have recently been carried out on reconstituted thin filaments, and support the conclusions based on negative staining (Xu et al. 1999). Tropomyosin is found in positions identical to those seen at high and low Ca^{2+} by negative staining. The only difference in the cryo-EM reconstruction is that the tropomyosin strand has a more circular cross-section than that seen by negative staining, and a radial position more consistent with that predicted by X-ray diffraction data (Poole et al. 1994, 1995).

2.6 Recent Criticism of the Steric Blocking Model

Despite the simplicity of the steric blocking model, and the structural support it has received from powerful electron microscopic, image processing and X-ray diffraction data, one argument against it remains to be answered. This concerns the identity of the protein strand interpreted as tropomyosin. This question arises because the interpretation of the different components in reconstructions is based primarily on shape. Thus, proteins with similar shapes could be confused. While the identity of the globular subunits as actin is not in doubt (owing to the excellent fit with the actin atomic structure and to comparison with reconstructions of pure F-actin), the identity of the elongated strand as tropomyosin has recently been questioned (Lehrer 1994; Squire and Morris 1998). The helical reconstruction procedure as used in these studies treats all densities associated with actin as if they were identical on each monomer along the filament. Other components of thin filaments might thus potentially contribute to the elongated strand, e.g. nebulin and the subunits of troponin, especially the elongated TnT molecule, which is half the length of tropomyosin, and generally considered to run parallel to it (Flicker et al. 1982; Cabral-Lilly et al. 1997). We showed earlier that nebulin does not contribute to the strand. The possibility that troponin may contribute remains.

Based on the observation that the *radial position* of tropomyosin hardly changes with Ca^{2+} (Mendelson et al. 1991; Popp and Maéda 1993), it has been suggested that the Ca^{2+}-induced structural change in the thin filament may be due mainly to movement of TnI, and that tropomyosin movement occurs only when myosin heads bind to actin (Lehrer 1994; Lehrer and Geeves 1998). A related suggestion, arising from reanalysis of X-ray diffraction and EM data, is that movement of the whole troponin complex, without movement of tropomyosin itself, could explain the changes in intensity of the X-ray diffraction pattern of activated muscle and the Ca^{2+}-induced strand movement observed in EM reconstructions (Squire and Morris 1998).

While these arguments have not yet been refuted, there are strong reasons to believe that the strand is due mainly to tropomyosin and that the Ca^{2+}-induced changes are due primarily to its movement:

1. The arguments depend on the fact that the helical reconstruction procedure can "average" discontinuous structures lying on the actin helix (e.g. troponin), potentially generating or contributing to *apparently continuous* ones. However, the strand observed in the reconstructions is also *directly* visualized (without any image processing) in some of the original images, where it is seen to be continuous over several long-pitch actin repeats (Fig. 1c). Because troponin occurs *discretely* along tropomyosin and is only half its length (Flicker et al. 1982), neither it nor its subunits can explain direct visualization of a continuous strand. The only component that can do so is tropomyosin.

2. Much of the data from the more globular parts of troponin are excluded from the reconstructions (by omitting "meridional" information). While extended regions of troponin are averaged over seven actins, only TnT is sufficiently long to be a factor, and it could appear in reconstructions with no more than half weight (being only half the length and mass of tropomyosin). Thus the contribution of troponin to the reconstructions is likely to be small.

3. This is strongly supported by the fact that a strand of protein density indistinguishable from that seen in thin filament reconstructions *containing* troponin is also seen in reconstructions of actin-tropomyosin filaments *lacking* troponin. Thus there is no reason to invoke any additional proteins to explain the density interpreted as tropomyosin.

4. Such strands can be seen (in the absence of any troponin) in both ON *and* OFF positions, the position depending on whether the tropomyosin is from skeletal or smooth muscle (see later). Thus strand visualization in ON *and* OFF positions can be accounted for by tropomyosin alone.

5. There is no evidence that TnI has an axially oriented, elongated structure that could account for the continuous strand (as required by one of the arguments (Lehrer 1994)) even if it were averaged in reconstructions. Such an elongated structure for TnI is not apparent in EM images of Tn (Flicker et al. 1982).

6. Although tropomyosin stays at the same radius at high and low Ca^{2+} (Mendelson et al. 1991; Popp and Maéda 1993), this does not mean that it does

not *move* over actin, as has been suggested (Lehrer 1994). The negative stain and cryo-EM results discussed above show that tropomyosin moves *azimuthally* by 2 nm in response to Ca^{2+}, while remaining in contact with the relatively flat face of actin at a constant radius of 3.8 nm.

7. Changes in the stoichiometry of Tn (and hence the ratio of Tn/Tm; Rosol et al. 1998) or in the molecular mass of troponin (Lehman et al. 1994, 1995) have no apparent impact on strand density in reconstructions.

2.7 Mechanistic Insights into Regulation Using Mutant Tropomyosin

Structural and functional studies have been carried out on thin filaments reconstituted using a bacterially expressed internal deletion tropomyosin, D234, in which actin-binding pseudorepeats 2, 3 and 4 are missing. This tropomyosin retains regions involved in end-to-end contact and troponin binding, but spans only four instead of the normal seven actin subunits (Landis et al. 1997). In 3-D reconstructions of filaments reconstituted from actin, D234 and troponin, tropomyosin is seen in the same high and low Ca^{2+} positions found with native tropomyosin (Rosol et al. 1998). Surprisingly, however, these filaments are functionally in the OFF state in both high and low Ca^{2+} (i.e., ATPase, tension and *in vitro* motility are inhibited even at high Ca^{2+}). This suggests that although most of the myosin binding site on actin is uncovered at high Ca^{2+}, D234 tropomyosin cannot be displaced by weakly attached, cycling crossbridges to the position that *fully* exposes the entire myosin binding site. The sub-stoichiometric binding of NEM-modified S1, which binds strongly to thin filaments even in the presence of ATP, overcomes the inhibition and switches the filaments ON, activating ATPase. These experiments add strength to the view that the binding of myosin heads to actin is an integral step of full switching ON of the thin filament (Rosol et al. 1998).

2.8 The Structure of Thin Filaments During Active Sliding with Myosin Filaments

The studies described to this point have been carried out on native thin filaments isolated from muscle, or on filaments reconstituted from purified components. A closer approach to the in vivo situation can be made by observing the structural changes in thin filaments interacting with thick filaments in unpurified muscle homogenates. Two questions have been studied using this approach: what is the position of tropomyosin? and where on actin do myosin heads bind in relaxed, rigor and actively sliding states (Craig and Lehman 1999)?

Under relaxing conditions (ATP, low Ca^{2+}), actin and myosin filaments occasionally run parallel to and in contact with each other, presumably reflecting weak, transient interaction of crossbridges with thin filaments.

Reconstructions of such actin filaments reveal tropomyosin in the same OFF position (the inner edge of the outer domain) observed with purified filaments (cf. Fig. 3a), blocking strong myosin binding, but not interfering with possible transient, weak cross-bridge binding on the periphery of subdomain-1. In muscle homogenates that have been activated by Ca^{2+}, thick and thin filaments are much more frequently in contact with each other, joined by readily visible, irregularly distributed, disordered cross-bridges. In reconstructions, tropomyosin is seen in the same position as that in high Ca^{2+} thin filaments not interacting with myosin heads (cf. Fig. 3b). In addition, extra density, not found in purified thin filaments, is seen on the periphery of the outer domain of actin, presumably representing the actin-binding region of active, cycling, weakly attached cross-bridges (Fig. 2f). In muscle homogenates that have been depleted of ATP, parallel actin and myosin filaments are commonly joined by regular, angled cross-bridges, characteristic of the rigor state. Reconstructions of such thin filaments show tropomyosin density in the position that fully exposes the myosin binding site, and cross-bridge density is seen in the same position as that in filaments decorated with S1 in the absence of ATP (cf. Fig. 3c). These results are the first to show directly the position of myosin heads on actin during early stages of the cross-bridge cycle, and support the view that initial contact occurs not at the strong binding site blocked by tropomyosin, but at the charged periphery of actin.

3 Regulation in Smooth Muscle

In vertebrate smooth muscle, troponin is absent from the thin filaments and the primary switch regulating actin-myosin interaction is phosphorylation of the regulatory light chains on the myosin heads (reviewed by Kamm and Stull 1985; Sellers and Adelstein 1986). Recent studies have suggested that this switching may be fine-tuned by the actin-binding proteins caldesmon and calponin (present along with tropomyosin in smooth muscle thin filaments but absent from skeletal muscle), leading to the graded contractions critical to the function of smooth muscle. Both proteins have been characterized by biochemical and physiological studies, and both have been shown to inhibit actomyosin ATPase and filament sliding in vitro (reviewed by Marston and Huber 1996; Gimona and Small 1996). However, the differing conclusions reached from these studies have not produced agreement on the roles played by these proteins in vivo. These uncertainties are compounded by the fact that smooth muscle apparently has two structural domains, a contractile and a cytoskeletal domain, each associated with specific proteins. The contractile domain contains myosin and smooth muscle α- or γ-actin, while the cytoskeletal domain contains cytoplasmic (β-) actin, desmin and filamin. Caldesmon appears to be confined to the contractile domain, while calponin is primarily in the cytoskeletal domain, but may also be in the contractile domain (ibid). Here we briefly review recent ultrastructural observations on the molecular organization of caldesmon and calponin on smooth muscle thin filaments that place constraints on possible models for their function.

3.1 Caldesmon

Caldesmon is an extremely long (50–80 nm, depending on its source), narrow (1.7 nm), flexible protein that has been implicated in modulating the state of assembly of the cytoskeleton in nonmuscle cells and contraction in smooth muscle (see Sobue and Sellers 1991; Marston and Huber 1996 for reviews). Immuno-EM shows that it is axially arranged along the thin filament, its position being determined by tropomyosin, and probably spans 14 actin monomers (Lehman et al. 1989; Moody et al. 1990), consistent with its stoichiometry on actin.

Reconstructions of thin filaments in the absence of caldesmon show that smooth muscle tropomyosin lies on the outer domain of actin, covering the strong myosin binding site (Fig. 2g,i). When caldesmon is present it constrains tropomyosin *away* from this site, on to the inner domain of actin (Vibert et al. 1993; Hodgkinson et al. 1997a; Fig. 2h,j). This is a surprising result, as activation of myosin ATPase is *inhibited* when caldesmon is present on thin filaments, and tropomyosin in this state is in the same position as the ON state of skeletal filaments, not in a position where it would block the strong myosin binding site. How, then, does caldesmon inhibit actin-myosin ATPase? Careful examination of reconstructions reveals the position of caldesmon itself, on the extreme outer edge of actin subdomains 1 and 2, not in obvious contact with tropomyosin on the inner domain of actin (Lehman et al. 1997). Fitting the reconstructions to the atomic model of F-actin shows that caldesmon appears to cover potential *weak* sites of myosin interaction on the periphery of subdomain-1, while, together with tropomyosin, it flanks strong sites of myosin interaction without covering them (Fig. 3d). This position supports the view that caldesmon may inhibit actin-myosin interaction by blocking weak, initial myosin binding, as suggested by ATPase and force measurements (e.g. Chalovich et al. 1987; Brenner et al. 1991). These observations imply that if caldesmon modulates actin-myosin interaction in vivo, it does so by a different mechanism from that of troponin in striated muscle thin filaments, where it is the *strong* binding site that is blocked, in this case by *tropomyosin*.

3.2 Calponin

Calponin is a 32-kDa actin-binding protein found in substantial quantities in smooth muscle (reviewed by Gimona and Small 1996). In vitro, it inhibits actomyosin ATPase and actin filament sliding on myosin (Haeberle 1994), while in intact cell models it may affect contractility and force maintenance (Itoh et al. 1994; Malmqvist et al. 1997), a characteristic of smooth muscle in which tension is maintained at low energy cost (Hai and Murphy 1989). While calponin *might* function in vivo to modulate actin-myosin interaction, it has also been suggested that calponin could have a purely cytoskeletal function, linking filaments to each other or to signal transduction pathways (Mabuchi et al. 1997; Menice et al. 1997). Calponin is currently of intense interest because conserved actin-binding modules known as "calponin homology (CH) domains" (Castresana and Saraste 1995) have been found in a variety of cytoskeletal actin-binding proteins. Some of these proteins (e.g. dystrophin, α-actinin, filamin) either cross-link actin filaments to each other or to other cytoskeletal elements. Others (e.g. Vav, IQGAP) modulate the binding of signaling proteins to the cytoskeleton (ibid.). Calponin may represent the archetypal mammalian CH-domain protein.

In reconstructions of actin-tropomyosin-calponin filaments, calponin density occurs peripherally on actin, mainly over subdomain-2, connecting axially adjacent actin monomers by the "upper" and "lower" edges of subdomains-1 respectively. Tropomyosin is displaced from subdomain-1 towards subdomain-3, exposing strong myosin-binding sites that were previously covered (Hodgkinson et al. 1997b). This structural effect (similar to that caused by caldesmon) is unlike that of troponin and therefore inhibition of actomyosin ATPase by calponin, like caldesmon, is different from that by troponin. When the atomic model of F-actin is fitted to the reconstructions, calponin is seen to contact actin near the N-terminus and at residues 349–352 close to the C-terminus of subdomain-1 on one monomer and residues 92–95 of subdomain-1 on the axially neighboring actin (Fig. 3e). These positions are consensus binding sites for a number of actin-associated proteins (McGough 1998), and are also near to sites of weak myosin interaction. Calponin did not appear to block strong myosin binding sites on actin. Its position could explain its ability to inhibit weak binding of dephosphorylated myosin heads to actin and thus act as a potential regulator of the "latch" state (Malmqvist et al. 1997). The location of calponin may also allow it to directly compete or interact with a number of actin-binding proteins (e.g. filamin, α-actinin), consistent with the view that calponin, alternatively or additionally to its modulating actin-myosin interaction, may be involved in modeling the cytoskeleton and/or cellular signaling transduction pathways (see Hodgkinson et al. 1997b). Further atomic level insights into the interaction of CH domains with actin are now possible by fitting the atomic structure of the CH-domain to the atomic model of the actin filament (Hanein et al. 1998).

4 Conclusions

The following general conclusions can be drawn from the preceding structural studies:

1. Muscle thin filaments are constructed from an arrangement of actin subunits following two long pitch helices that twist around each other.

2. Tropomyosin molecules form head-tail polymers, one associated with each of the two long-pitch actin helices.

3. In the absence of other actin-binding proteins, tropomyosin binds to actin in one of two discrete positions ("equilibrium" positions), which depend on the source of tropomyosin. Striated muscle tropomyosin binds to the inner domain of actin while smooth muscle tropomyosin binds to the outer domain.

4. Tropomyosin position is critically affected by additional proteins present in native filaments.

5. In striated muscle thin filaments, in the absence of Ca^{2+}, troponin constrains tropomyosin to lie away from its equilibrium position, on the outer domain of actin, where it physically blocks the strong binding of myosin heads to actin (inhibiting ATPase and cross-bridge cycling) but does not interfere with weak electrostatic binding.

6. The binding of Ca^{2+} by troponin allows tropomyosin to return to its equilibrium position on the inner domain, exposing most of the strong site and removing the inhibition.

7. The binding of myosin heads causes a further movement of tropomyosin, allowing additional heads to bind, cooperatively activating contraction.

8. In smooth muscle, whatever the mechanism of the putative modulation of actin-myosin interaction by caldesmon or calponin, it differs from the troponin mechanism in striated muscle. One possibility appears to be that caldesmon or calponin themselves inhibit the initial binding of myosin heads to actin.

5 Future Directions

The steric blocking mechanism appears to be on a solid structural foundation, where uncertainties involving interpretation of X-ray diffraction data and electron microscopic images have been mostly removed. Future studies should be aimed at: (1) demonstrating beyond any remaining doubt that the protein strand observed in reconstructions is indeed tropomyosin, by defining the TnI, TnT and TnC contributions, and (2) determining the 3-D organization of troponin or its subunits on the thin filament, and the structural changes that occur on binding Ca^{2+}.

Acknowledgements. This publication was made possible by grant numbers AR34711, HL47530, HL62468 and HL36153 from the National Institutes of Health. Its contents are solely the responsibility of the authors and do not necessarily represent the official views of the NIH.

References

al-Khayat HA, Yagi N, Squire JM (1995) Structural changes in actin-tropomyosin during muscle regulation: computer modelling of low-angle X-ray diffraction data. J Mol Biol 252: 611–632

Bremel RD, Weber A (1972) Cooperation within actin filament in vertebrate skeletal muscle. Nature 238:97–101

Brenner B, Yu LC, Chalovich JM (1991) Parallel inhibition of active force and relaxed fiber stiffness in skeletal muscle by caldesmon: implications for the pathway to force generation. Proc Natl Acad Sci USA 88:5739–5743

Cabral-Lilly D, Tobacman LS, Mehegan JP, Cohen C (1997) Molecular polarity in tropomyosin-troponin T co-crystals. Biophys J 73:1763–1770

Castresana J, Saraste M (1995) Does Vav bind to F-actin through a CH domain? FEBS Lett 374:149–151

Chalovich JM (1992) Actin mediated regulation of muscle contraction. Pharmacol Ther 55:95–148

Chalovich JM, Cornelius P, Benson CE (1987) Caldesmon inhibits skeletal actomyosin subfragment-1 ATPase activity and the binding of myosin subfragment-1 to actin. J Biol Chem 262: 5711–5716

Cohen C (1975) The protein switch of muscle contraction. Sci Am 233:36–45

Cohen C, Vibert PJ (1987) Actin filaments: images and models. In: Squire JM, Vibert PJ (eds) Fibrous protein structure. Academic Press, London, pp 283–306

Craig R, Lehman W (1999) Crossbridge and tropomyosin positions observed in native, interacting thick and thin filaments. Biophys J 76:A274

DePue RH, Rice RV (1965) F-actin is a right-handed helix. J Mol Biol 12:302–303

Ebashi S, Endo M, Ohtsuki I (1969) Control of muscle contraction. Q Rev Biophys 2:351–384

El-Saleh SC, Warber KD, Potter JD (1986) The role of tropomyosin-troponin in the regulation of skeletal muscle contraction. J Muscle Res Cell Motil 7:387–404

Flicker PF, Phillips, GN, Cohen, C (1982) Troponin and its interactions with tropomyosin. J Mol Biol 162:495–501

Gillis JM, O'Brien EJ (1975) The effect of calcium ions on the structure of reconstituted muscle thin filaments. J Mol Biol 99:445–459

Gimona M, Small JV (1996) Calponin. In: Bárány M (ed) Biochemistry of smooth muscle contraction. Academic Press, San Diego, pp 91–103

Hai C-M, Murphy RA (1989) Ca^{2+}, crossbridge phosphorylation, and contraction. Annu Rev Physiol 51:285–298

Haeberle JR (1994) Calponin decreases the rate of cross-bridge cycling and increases maximum force production by smooth muscle myosin in an in vitro motility assay. J Biol Chem 269: 12424–12431

Hanein D, Volkmann N, Goldsmith S, Michon AM, Lehman W, Craig R, DeRosier D, Almo S, Matsudaira P (1998) An atomic model of fimbrin binding to F-actin and its implications for filament crosslinking and regulation. Nature Struct Biol 5:787–792

Hanson J (1973) Evidence from electron microscopic studies on actin paracrystals concerning the origin of the cross-striation in the thin filaments of vertebrate skeletal muscle. Proc R Soc Lond B 183:39–58

Hanson J, Lowy J (1963) The structure of F-actin and of actin isolated from muscle. J Mol Biol 6:46–60

Hanson J, Lednev V, O'Brien EJ, Bennett PM (1973) Structure of the actin-containing filaments in vertebrate skeletal muscle. Cold Spring Harbor Symp Quant Biol 37:311–318

Haselgrove JC (1973) X-ray evidence for a conformational change in the actin-containing filaments of vertebrate striated muscle. Cold Spring Harbor Symp Quant Biol 37:341–352

Hodgkinson JL, Marston SB, Craig R, Vibert P, Lehman W (1997a) Three-dimensional image reconstruction of reconstituted smooth muscle thin filaments: effects of caldesmon. Biophys J 72:2398–2404

Hodgkinson JL, EL-Mezgueldi M, Craig R, Vibert P, Marston SB, Lehman W (1997b) 3-D image reconstruction of reconstituted smooth muscle thin filaments containing calponin: visualization of interactions between F-actin and calponin. J Mol Biol 273:150–159

Holmes KC, Popp D, Gebhard W, Kabsch W (1990) Atomic model of the actin filament. Nature 347:44–49

Huxley HE (1963) Electron microscope studies on the structure of natural and synthetic protein filaments from striated muscle. J Mol Biol 7:281–308
Huxley HE (1973) Structural changes in the actin- and myosin-containing filaments during contraction. Cold Spring Harbor Symp Quant Biol 37:361–376
Itoh T, Suzuki S, Suzuki A, Nakamura F, Naka M, Tanaka T (1994) Effects of exogenously applied calponin on Ca^{2+}-regulated force in skinned smooth muscle of the rabbit mesenteric artery. Pfluegers Arch 427:301–308
Kabsch W, Mannherz HG, Suck D, Pai EF, Holmes KC (1990) Atomic structure of the actin: DNase I complex. Nature 347:37–44
Kamm KE, Stull JT (1985) The function of myosin and myosin light chain kinase phosphorylation in smooth muscle. Annu Rev Pharmacol Toxicol 25:593–620
Kress M, Huxley HE, Faruqi AR, Hendrix J (1986) Structural changes during activation of frog muscle studied by time-resolved X-ray diffraction. J Mol Biol 188:325–342
Landis CA, Bobkova A, Homsher E, Tobacman LS (1997) The active state of the thin filament is destabilized by an internal deletion in tropomyosin. J Biol Chem 272:14051–14056
Lehman W, Craig R, Lui J, Moody C (1989) Caldesmon and the structure of smooth muscle thin filaments: immunolocalization of caldesmon on thin filaments. J Muscle Res Cell Motil 10: 101–112
Lehman W, Craig R, Vibert P (1994) Ca^{2+}-induced tropomyosin movement in *Limulus* thin filaments revealed by three-dimensional reconstruction. Nature 368:65–67
Lehman W, Vibert P, Uman P, Craig R (1995) Steric-blocking by tropomyosin visualized in relaxed vertebrate muscle thin filaments. J Mol Biol 251:191–196
Lehman W, Vibert P, Craig R (1997) Visualization of caldesmon on smooth muscle thin filaments. J Mol Biol 274:310–317
Lehrer SS (1994) The regulatory switch of the muscle thin filament: Ca^{2+} or myosin heads? J Muscle Res Cell Motil 15:232–236
Lehrer SS, Geeves MA (1998) The muscle thin filament as a classical cooperative/allosteric regulatory system. J Mol Biol 277:1081–1089
Lorenz M, Popp D, Holmes K (1993) Refinement of the F-actin model against X-ray fiber diffraction data by the use of a directed mutation algorithm. J Mol Biol 234:826–836
Mabuchi K, Li B, Ip W, Tao T (1997) Association of calponin with desmin intermediate filaments. J Biol Chem 272:22662–22666
Malmqvist U, Trybus KM, Yagi S, Carmichael J, Fay FS (1997) Slow cycling of unphosphorylated myosin is inhibited by calponin thus keeping smooth muscle relaxed. Proc Natl Acad Sci USA 94:7655–7660
Marston SB, Huber PAJ (1996) Caldesmon. In: Bárány M (ed) Biochemistry of smooth muscle contraction. Academic Press, San Diego, pp 77–90
McGough A (1998) F-actin-binding proteins. Curr Opin Struct Biol 8:166–176
McLachlan AD, Stewart M (1976) The 14-fold periodicity in α-tropomyosin and the interaction with actin. J Mol Biol 103:271–298
McKillop DFA, Geeves MA (1993) Regulation of the interaction between actin and myosin subfragment 1: evidence for three states of the thin filament. Biophys J 65:693–701
Mendelson RA, Stone DB, Timmins PA, Johnston ER (1991) Results of preliminary studies of tropomyosin radial movement using neutron scattering. Biophys J 59:219a
Menice CB, Hulvershorn J, Adam LP, Wang C-LA, Morgan KG (1997) Calponin and mitogen-activated protein kinase signaling in differentiated vascular smooth muscle. J Biol Chem 272:25157–25161
Milligan RA, Flicker PF (1987) Structural relationship of actin, myosin, and tropomyosin revealed by cryo-electron microscopy. J Cell Biol 105:29–39
Milligan RA, Whittaker M, Safer D (1990) Molecular structure of F-actin and location of surface binding sites. Nature 348:217–221
Moncman CL, Wang K (1995) Nebulette: a 107 kD nebulin-like protein in cardiac muscle. Cell Motil Cytoskeleton 32:205–225

Moody C, Lehman W, Craig R (1990) Caldesmon and the structure of smooth muscle thin filaments: electron microscopy of isolated thin filaments. J Muscle Res Cell Motil 11:176–185
Moore PB, Huxley HE, DeRosier DJ (1970) Three-dimensional reconstruction of F-actin, thin filaments and decorated thin filaments. J Mol Biol 50:279–295
O'Brien EJ, Dickens MJ (1983) Actin and thin filaments. In: Harris JR (ed) Electron microscopy of proteins. Academic Press, London, pp 1–95
Ohtsuki I, Masaki T, Nonomura Y, Ebashi S (1967) Periodic distribution of troponin along the thin filament. J Biochem (Tokyo) 61:817–819
Parry DAD, Squire, JM (1973) Structural role of tropomyosin in muscle regulation: analysis of the X-ray diffraction patterns from relaxed and contracting muscles. J Mol Biol 75:33–55
Phillips GN, Fillers JP, Cohen C (1986) Tropomyosin crystal structure and muscle regulation. J Mol Biol 192:111–131
Poole KJV, Lorenz M, Evans G, Rosenbaum G, Holmes KC (1994) The effect of calcium on the regulated thin filament structure. Biophys J 66:A347
Poole KJV, Evans G, Rosenbaum G, Lorenz M, Holmes KC (1995) The effect of crossbridges on the calcium sensitivity of the structural change of the regulated thin filament. Biophys J 68:A365
Popp D, Maéda Y (1993) Calcium ions and the structure of muscle actin filament. An X-ray diffraction study. J Mol Biol 229:279–285
Rayment I, Holden HM, Whittaker, M, Yohn CB, Lorenz M, Holmes KC, Milligan RA (1993) Structure of the actin-myosin complex and its implications for muscle contraction. Science 261: 58–65
Rome EM, Hirabayashi T, Perry SV (1973) X-ray diffraction of muscle labelled with antibody to troponin-C. Nature 244:154–155
Rosol M, Landis C, Lehman W, Craig R, Tobacman L (1998) 3-D reconstruction of inhibited thin filaments containing mutant tropomyosin. Biophys J 74:A143
Selby CC, Bear RS (1956) The structure of actin-rich filaments of muscles according to X-ray diffraction. J Biophys Biochem Cytol 2:71–85
Sellers JR, Adelstein RS (1986) Regulation of contractile activity. In: Boyer PD, Krebs EG (eds) The enzymes. Academic Press, Orlando, pp 381–418
Seymour J, O'Brien EJ (1980) The position of tropomyosin in muscle thin filaments. Nature 283:680–682
Seymour J, O'Brien EJ (1985) Structure of myosin decorated actin filaments and natural thin filaments. J Muscle Res Cell Motil 6:725–756
Sobue K, Sellers JR (1991) Caldesmon, a novel regulatory protein in smooth muscle and non-muscle actomyosin systems. J Biol Chem 266:12115–12118
Spudich JA, Huxley HE, Finch JT (1972) Regulation of skeletal muscle contraction – structural studies of the interaction of the tropomyosin-troponin complex with actin. J Mol Biol 72: 619–632
Squire JM, Morris EP (1998) A new look at thin filament regulation in vertebrate skeletal muscle. FASEB J 12:761–771
Squire JM, Al-Khayat HS, Yagi N (1993) Muscle thin-filament structure and regulation actin subdomain movements and the tropomyosin shift modelled from low-angle X-ray diffraction. J Chem Soc 89:2717–2726
Taylor KA, Amos LA (1981) A new model for the geometry of the binding of myosin crossbridges to muscle thin filaments. J Mol Biol 147:297–324
Tobacman LS (1996) Thin filament-mediated regulation of cardiac contraction. Annu Rev Physiol 58:447–481
Trinick J (1994) Titin and nebulin: protein rulers in muscle? TIBS 19:405–409
Wakabayashi T, Huxley HE, Amos LA, Klug A (1975) Three-dimensional image reconstruction of actin-tropomyosin complex and actin-tropomyosin-troponin T-troponin I complex. J Mol Biol 93:477–497
Weber A, Murray JM (1973) Molecular control mechanisms in muscle contraction. Physiol Rev 53:612–673

White SP, Cohen C, Phillips GN (1987) Structure of co-crystals of tropomyosin and troponin. Nature 325:826–828

Vibert P, Craig R, Lehman W (1993) Three-dimensional reconstruction of caldesmon-containing smooth muscle thin filaments. J Cell Biol 123:313–321

Vibert P, Craig R, Lehman W (1997) Steric-model for activation of muscle thin filaments. J Mol Biol 266:8–14

Xu C, Craig R, Horowitz R, Lehman W (1999) High and low Ca^{2+} positions of tropomyosin revealed by cryo-electron microscopy of thin filaments. Biophys J 77:985–992

The Role of Troponin in the Ca^{2+}-Regulation of Skeletal Muscle Contraction

Danuta Szczesna and James D. Potter[1]

1 The Role of TnC and Its Interaction with TnI in the Regulation of Contraction

Vertebrate skeletal muscle contraction is activated by the binding of Ca^{2+} to the low affinity Ca^{2+}-specific (regulatory) sites of troponin C (TnC), the Ca^{2+}-binding subunit of troponin, which together with troponin I (TnI), troponin T (TnT) and tropomyosin (Tm) form the regulatory system of the contractile apparatus (Zot and Potter 1987; Moss 1992; Gergely 1998). TnC is necessary for force development in skinned muscle fibers and its selective extraction from muscle fibers results in a permanent relaxation at all Ca^{2+} concentrations. This process is fully reversible and reconstitution of the TnC-depleted fibers with exogenous TnC restores Ca^{2+}-dependent contraction (Zot and Potter 1982; Szczesna et al. 1996). The structure of TnC has been solved to atomic resolution by X-ray crystallography (Herzberg and James 1985, 1988; Houdusse et al. 1997). TnC consists of two globular domains corresponding to the NH_2- and COOH-termini and each domain contains two EF-hand divalent cation-binding sites. The low affinity sites of TnC, called the Ca^{2+}-specific sites (I and II) (Potter and Gergely 1975), are located in the NH_2-terminal domain of TnC and are separated from the COOH-terminal domain of TnC by a single nine-turn α-helix comprising helices D and E (Herzberg and James 1985, 1988). The COOH-terminal region of TnC contains two high affinity Ca^{2+}-binding sites designated as sites III and IV and referred to as the Ca^{2+}-Mg^{2+} sites (Potter and Gergely 1975). Sites I and II bind Ca^{2+} specifically with $K_{Ca2+} \cong 3 \times 10^5\,M^{-1}$, whereas sites III and IV bind Ca^{2+} with $K_{Ca2+} \cong 2 \times 10^7\,M^{-1}$ and Mg^{2+} with $K_{Mg2+} \cong 2 \times 10^3\,M^{-1}$ (Potter and Gergely 1975; Grabarek et al. 1992). Under physiological conditions, in relaxed muscle, sites III and IV of TnC are primarily occupied by Mg^{2+} and can become partially saturated with Ca^{2+} during contraction, depending on the time course of the $[Ca^{2+}]$ transient (Robertson et al. 1981). While the process of Mg^{2+} exchange for Ca^{2+} is much too slow (k_{off}~$8\,s^{-1}$) to trigger muscle contraction, the kinetics of Ca^{2+} binding to the NH_2-terminal domain (sites I and II) of TnC are coupled with the rate of muscle activation, implying that these sites are directly involved in muscle

[1] Department of Molecular and Cellular Pharmacology, University of Miami School of Medicine, 1600 N.W. 10th Ave., Miami, Florida 33136, USA

Results and Problems in Cell Differentiation, Vol. 36
D.D. Thomas, C.G. dos Remedios (Eds.): Molecular Interactions of Actin

regulation. Ca^{2+} binding at these sites leads to the structural change within the NH_2-terminal domain of TnC and the exposure of a hydrophobic pocket (Johnson et al. 1978; Herzberg et al. 1986, Strynadka et al. 1997; Houdusse et al. 1997; Gagne et al. 1994, 1995, 1997, 1998; She et al. 1998). Interestingly, a recent 4 Ca^{2+}-bound structure of recombinant rabbit TnC (Houdusse et al. 1997) confirmed the earlier predicted Ca^{2+}- bound conformation of the NH_2-terminal domain (Herzberg et al. 1986). A recent study from Moncrieffe et al. (1999a) indicates that the binding of Ca^{2+} to the NH_2-terminal domain of TnC (4 Ca^{2+} state) results in a more elongated structure of the molecule compared with the 2 Ca^{2+} state (sites III and IV saturated with Ca^{2+}).

Besides the two Ca^{2+}-specific regulatory sites flanked respectively by helices A, B and C, D, the NH_2-terminal domain of TnC contains a 14-residue N-helix which was shown to have a functional role (Gulati et al. 1993; Liu et al. 1994; Chandra et al. 1994a; Smith et al. 1994, 1999). The two globular Ca^{2+} binding domains of TnC are connected by a three-turn central helix (D/E). The recently resolved crystal structure of the 4 Ca^{2+}-bound TnC reveals that the central helix is more stable in the 2 Ca^{2+}-bound state (Ca^{2+}-bound in the high affinity, COOH-terminal domain sites) than when all four sites of TnC are saturated with Ca^{2+} (Houdusse et al. 1997). The structural requirements of this helical linker in TnC are not fully understood. Deletion of up to two turns (seven residues) did not greatly affect the regulatory function (Dobrowolski et al. 1991; Sheng et al. 1991), whereas deletion of more than seven residues showed dramatic loss of TnC function in a skinned skeletal muscle fiber system (Babu et al. 1993). In another study an 11-residue deletion mutant demonstrated defective interaction with TnI (Ramakrishnan and Hitchcock-DeGregori 1995). Among all central helix residues, the aspartic acid at the position 89 (D89) was shown to be crucial for Ca^{2+}-dependent thin filament activation irrespective of the helical environment (Ramakrishnan and Hitchcock-DeGregori 1996). It has been postulated that native length and structure of the central helix are optimal for normal regulatory function of TnC (Ramakrishnan and Hitchcock-DeGregori 1995), possibly for Ca^{2+} signaling and communication between the NH_2- and the COOH-terminal domains of TnC (Grabarek et al. 1986, 1992; Wang et al. 1993; Moncrieffe et al. 1999b). A number of studies have suggested a structural role for the high affinity Ca^{2+}-Mg^{2+} binding sites, III and IV, indicating that they are involved in maintaining the stability of the whole troponin complex in the thin filament, presumably through the interactions between TnC and the inhibitory subunit of troponin, TnI (Potter and Gergely 1975; Zot and Potter 1982). We have been studying the structural determinants of TnC that govern its regulatory (functional) and structural properties in muscle contraction with the use of skinned skeletal muscle fibers where steady state force and the Ca^{2+} sensitivity of force development are easily accessed. These fibers can be depleted of endogenous TnC and reconstituted with various TnC mutants or fragments. Using this system we have shown that various regions of TnC play different roles in the regulation of contraction (Szczesna et al. 1996). In another study, we investigated the functional role of the domains of TnC through the use of thrombin fragments of TnC reconstituted into skeletal muscle fibers (Francois et al. 1995). Four thrombin fragments containing the following Ca^{2+}-binding site(s): TH1 (residues 1–120) sites I–III; TH2 (121–159) site IV; TH3 (1–100) sites I and II; and TH4 (101–120) site III, were used. We have shown that the NH_2-terminal domain fragments (TH1 and TH3) of TnC maintained the regulatory properties of intact TnC. They were able to restore about 50–65% of force compared with unextracted fibers. Like intact TnC, TH1 was able to interact with the fibers in a Ca^{2+}-independent (Mg^{2+}-dependent) manner, indicating that at a minimum, Ca^{2+}-binding site III is required for this type of binding. The COOH-terminal domain fragments (TH2 and TH4) were mostly involved in the Ca^{2+}-Mg^{2+}-dependent interactions of TnC with the fibers. It appeared that they bound to two different binding sites in the fibers. The binding to one of the sites occurred only in the presence of Ca^{2+} and caused partial force restoration. This binding was abolished when Ca^{2+} was removed. The binding to the second site required Ca^{2+} initially, but was maintained in the presence of Mg^{2+}. This interaction partially blocked the rebinding of intact TnC to the fibers. These results indicated that site III or IV in these small fragments, when removed from the constraints of the parent protein, may assume conformations that allow them to function, to a certain extent, like both the regulatory sites (I and II) and the Ca^{2+}-Mg^{2+} sites (III and IV) of TnC. The role of the Ca^{2+}-specific and the high affinity Ca^{2+}-Mg^{2+} sites of TnC were studied by constructing four mutants and wild type (WTnC) of chicken skeletal TnC, with inactivated Ca^{2+} binding sites I and II (TnC1,2–), site III (TnC3–), site IV (TnC4–) and sites III and IV (TnC3,4–) (Szczesna et al. 1996). All Ca^{2+} binding site mutations were generated by replacing the Asp at the X-coordinating position of the Ca^{2+} binding loop with Ala. The binding of these mutated proteins to TnC-depleted skinned skeletal muscle fibers was investigated, as well as the rate of their dissociation from these fibers. The proteins were also tested for their ability to restore steady state force to TnC-depleted fibers. We found that although the NH_2-terminal mutant of TnC (TnC1,2–) bound to the TnC-depleted fibers (with a lower affinity than WTnC), it was unable to reactivate Ca^{2+}-dependent force. This supports earlier findings (Sheng et al. 1990) that the low affinity Ca^{2+} binding sites (I and II) in TnC are responsible for the Ca^{2+}-dependent activation of skeletal muscle contraction. All three COOH-terminal mutants of TnC bound to the TnC-depleted fibers had different rates of dissociation and could restore steady state force to the level of non-extracted fibers. Although both high affinity Ca^{2+} binding sites (III and IV) are important for binding to the fibers, site III appears to be the primary determinant for maintaining the structural stability of TnC in the thin filament. Moreover, our results suggest an interaction between the NH_2- and COOH-terminal domains of TnC (Grabarek et al. 1986, 1992; Wang et al. 1993), since alteration of sites I and II lowers the binding affinity of TnC to the fibers and mutations in sites III and IV affected the Ca^{2+} sensitivity of force development.

The interaction interface of the COOH-terminal domain of TnC containing the high affinity Ca^{2+}-Mg^{2+} sites, III and IV, has been shown to be located in the

NH_2-terminal region of TnI (residues 1–98) and also near the NH_2-terminal end of the inhibitory region of TnI containing residues 96–116. This so-called antiparallel fashion of TnC-TnI interaction is also true for the Ca^{2+}-dependent interaction of the NH_2-terminal regulatory domain of TnC with the COOH-terminal region of TnI, as well as with the COOH-terminal end of the TnI inhibitory peptide (Dalgarno et al. 1982; Leszyk et al. 1987, 1990, 1998; Farah et al. 1994; Kobayashi et al. 1995). Sheng et al. (1992) have revealed, utilizing a TnI deletion mutant missing the first 57 amino acid residues (TnI_{d57}), that there are at least three types of interaction between TnC and TnI: (1) one which depends on binding Ca^{2+} to the regulatory sites of TnC; (2) one which depends upon Mg^{2+} or Ca^{2+} binding to the high affinity Ca^{2+}-Mg^{2+} sites; and (3) metal independent interaction. The authors showed that the absence of the NH_2-terminal domain of TnI resulted in a loss of the Ca^{2+}-Mg^{2+} dependent interaction with TnC. Further, the authors have shown that TnI_{d57} was still able to maintain the Ca^{2+}-dependent regulation of the actomyosin ATPase activity, possibly through its interaction with the regulatory domain of TnC, indicating an antiparallel type of interaction between TnI and TnC (Potter et al. 1995).

Farah et al. 1994, have mapped the sites of TnC-TnI interaction using a series of recombinant TnI deletion mutants, TnI_{1-98}, TnI_{1-116}, TnI_{1-156}, $TnI_{120-182}$, $TnI_{103-182}$, as well as the NH_2- and COOH-domain regions of TnC. The authors confirmed an antiparallel arrangement of the TnC-TnI interaction and proposed a model for the thin filament regulation by troponin, refined later in Malnic et al. (1998). Basically, two classes of interactions within the troponin complex have been defined: (1) the structural interactions that occur between the COOH-terminal domain of TnC, the NH_2-domain of TnI, and the COOH-terminal domain of TnT, and (2) the regulatory interactions occurring between the NH_2-domain of TnC, the COOH-terminal domains of TnI and TnT, and actin-Tm (Malnic et al. 1998).

In a recent paper of Ramos (1999), a series of gradual deletion mutants of the COOH-terminal region of TnI were constructed to demonstrate that the whole COOH-terminus of TnI is necessary for full regulatory activity. The region between residues 137–144 of TnI was shown to interact with TnC and the 98–129 region of TnI to modulate the affinity of TnC for Ca^{2+}. Residues 166–182 were shown to be involved in the binding of TnI to thin filament (Ramos 1999). Luo et al. (1999) have shown that the region of Met121 of TnI stays in close contact with the NH_2-terminal hydrophobic pocket of TnC (residues 48 and 82) and that the thin filament segment containing TnI (Met121) moves away slightly from the hydrophobic patch in the absence of Ca^{2+}.

The inhibitory function of TnI has been studied in many different ways using synthetic peptides, as well as the proteolytic and recombinant fragments of TnI. The cyanogen bromide fragment of TnI (CN4), consisting of the amino acid residues 96–116, was shown to possess all of the inhibitory properties of TnI (Syska 1976). Studies with synthetic peptides have shown that residues 105–114 are the minimum sequence necessary for the inhibition of actomyosin

ATPase activity and TnC binding (Talbot and Hodges 1979; Van Eyk et al. 1991), however the CN4 fragment (TnI_{96-116}) has been shown to have an eight-fold higher affinity for TnC compared with $TnI_{104-115}$ (Chandra et al. 1994b). Moreover, reconstituted TnI fragments containing the inhibitory region of TnI, plus either the NH_2- (TnI_{1-116}) or COOH- (TnI_{96-148}) terminal regions of TnI were shown to be responsible for either maintaining the maximal level of acto-myosin ATPase activity or maintaining the Ca^{2+} dependence of ATPase, respectively (Van Eyk et al. 1997). Residues 116–126 of TnI were found to be critical for allowing TnC to fully neutralize the acto-S1-Tm inhibition caused by various TnI peptides (Tripet et al. 1997). The regulatory complex composed of TnT, TnC and the fragment of TnI (96–148) displayed the full regulatory properties of troponin, suggesting that TnI peptide 96–148 constitutes the major sequence of TnI responsible for an inhibitory activity similar to that of intact TnI (Tripet et al. 1997; Van Eyk et al. 1997). It has been demonstrated that the residues 96–116 of TnI are primarily responsible for binding to the COOH-domain of TnC, and residues 117–148 of TnI to the NH_2-terminal domain of TnC (Pearlstone et al. 1997; Van Eyk et al. 1997). Based upon these experiments, the inhibitory region of TnI was extended to residues 96–148 (McKay 1997, 1998; Pearlstone et al. 1997; Tripet et al. 1997; Van Eyk et al. 1997). Studies with TnI mutants (Szczesna et al. 1999) suggest that the COOH-terminal region of the 96–116 sequence of TnI (residues 105–115) is much more critical for the inhibitory activity of TnI in skinned skeletal muscle fibers than the NH_2-terminal region consisting of residues 95–106. The structure of the inhibitory peptide (96–115) was presented in a recent study of Hernandez et al. (1999). Utilizing CD and NMR spectroscopy, the authors demonstrated a predominantly extended conformation of the free TnI peptide with no significant change on binding to TnC. Analysis of the CD spectrum of the free peptide indicated only about 3% helix and no change in total helix content upon complex formation with TnC (Hernandez et al. 1999). This is opposite to that found by Campbell and Sykes (1991), who demonstrated that a smaller inhibitory peptide of TnI (104–115) forms two helical regions (104–108 and 112–115) upon binding TnC.

Molecular interactions between troponin subunits have been studied with high-resolution techniques such as X-ray crystallography, NMR or neutron scattering. A model of the TnI•TnC complex has been proposed based upon the low-angle scattering data (Olah et al. 1994) where TnI is an elongated, Z-shaped molecule twisted around TnC. In the ternary complex, together with TnC and TnT, and in the presence of regulatory concentrations of Ca^{2+}, the TnI molecule was much less elongated than in the binary TnI•TnC complex, based on the same neutron scattering method (Stone et al. 1998). Interestingly, the authors observed a large-scale in situ movement of mass within TnI in response to Ca^{2+}-binding to the regulatory sites of TnC. Since the gross conformational changes occurred in TnI and not in TnC, the TnI molecule was identified as part of the Ca^{2+} sensitive molecular switch (Stone et al. 1998).

A recent crystal structure of TnC, complexed with a TnI peptide comprising the first 47 amino acids of TnI, has shed more light on the atomic interactions between TnC and TnI (Vassylyev et al. 1998). This structure of TnC•TnI_{1-47} confirmed the antiparallel type of interaction between the COOH-terminal domain of TnC and the NH_2-terminal region of TnI, and also revealed a more compact organization of TnC in this complex compared with the free state of TnC (Herzberg et al. 1986). Based on this newly resolved TnC-TnI_{1-47} structure, the authors modeled atomic interactions between the inhibitory region of TnI_{96-127} (TnI_{reg}) and the NH_2-lobe of TnC, suggesting that this region of TnI acts as a Ca^{2+}-sensitive switch in muscle contraction, which moves between actin-Tm and the hydrophobic pocket of the NH_2-terminal domain of TnC (Vassylyev et al. 1998). A similar mechanism was proposed by McKay et al.1998, based upon the NMR structure of the regulatory, NH_2-terminal domain of TnC and the elongated inhibitory peptide of TnI consisting of residues 96–148. The authors demonstrated that this region of TnI interacts with the NH_2-terminal domain of TnC in a Ca^{2+}-sensitive manner and that this regulatory domain of TnC does not undergo a major structural change when binding to TnI_{96-148}. Binding of Ca^{2+} to the regulatory sites of TnC sufficiently exposes the hydrophobic pocket for the interaction of TnC with TnI and no further significant change in TnC conformation could be seen upon binding of TnI_{96-148} (McKay et al. 1998). In fact, it was shown earlier (Herzberg et al. 1986; Gagne et al. 1994, 1995) and most recently (Gagne et al. 1997, 1998; Houdusse et al. 1997; Strynadka et al. 1997; She et al. 1998) that the binding of Ca^{2+} to the regulatory sites of TnC causes a significant change in TnC conformation, opening its NH_2- terminal pocket for the interaction with TnI. This is in agreement with recent data from Moncrieffe et al. (1999a), where 4 Ca^{2+} TnC did not change its elongated shape upon binding a TnI peptide consisting of residues 96–116 (TnI_{96-116}), based on measurements of time-resolved fluorescence anisotropy and sedimentation velocity. However, a large change in shape was observed between the 2 Ca^{2+} TnC-TnI_{96-116} and 4 Ca^{2+} TnC-TnI_{96-116} complexes, where the former changed from a spherical shape to an elongated prolate ellipsoid in the latter (Moncrieffe et al. 1999a).

2 The Role of TnT and Its Interaction with TnC in the Regulation of Contraction

Equally important is understanding the role of the third Tn subunit, troponin T (TnT). Previous studies have suggested that a primary role for TnT is to interact with and anchor the TnI•TnC complex to the thin filaments through the interaction of TnT with both TnI and Tm (for reviews see Zot and Potter 1987). Numerous studies have demonstrated that when complexed with TnI, TnC and Tm, TnT confers Ca^{2+} sensitivity to the actomyosin ATPase (Greaser and Gergely 1971; Potter et al. 1995; Malnic et al. 1998; Perry 1998 and references therein). A high-resolution structure of TnT or its proteolytic fragments is not yet available. However, X-ray studies of Tm and TnT co-crystals at 17-Å resolution suggest that TnT is quite disordered (White et al. 1987). Electron microscopic studies on the isolated molecule show that TnT is 185 ± 25 Å long and about 20 Å wide (Flicker et al. 1982), which is in agreement with earlier antibody work on thin filaments (Ohtsuki 1979). Structural studies have shown that TnT extends along the COOH-terminal third of Tm and the NH_2-terminus of TnT binds to the head-to-tail overlap region of two adjacent Tm molecules (Brisson 1986; Ohtsuki et al. 1986; White et al. 1987).

Within a single muscle type (cardiac or skeletal), TnT is unique in that there are many more isoforms compared with TnC or TnI (Abe et al. 1986; Smillie et al. 1988; Briggs et al. 1994). In skeletal muscle, TnT is generated from ten to potentially as many as 64 different mRNAs. The capacity of these various TnT isoforms to associate in alternate combinations may play a role in determining the physiological regulation of muscle function or assembly. Thus it is important to understand the significance of the multiple isoforms of TnT in the regulation of muscle contraction. The sequence variations among the isoforms in a given species are confined to a region of approximately 30 amino acids (aa) near the NH_2-terminus and a shorter sequence near the COOH-terminus of TnT (Breitbart et al. 1985; Briggs et al. 1989). This is shown in Fig. 1A, B for rabbit skeletal muscle TnT that illustrates the three major TnT isoforms (TnT_{1f}, TnT_{2f}, and TnT_{3f}) that are expressed in rabbit muscle. These isoforms appear to arise from alternative splicing of primary transcripts of a single TnT gene (Breitbart et al. 1985; Medford et al. 1984; Cooper and Ordahl 1985). The Nadal-Ginard group (Medford et al. 1984) has demonstrated that in rat, a minimum of ten NH_2-terminal isoforms of TnT are present as a result of complicated differential splicing of five 5′ exons of the TnT gene, whereas only two COOH-terminal isoforms, α and β, are generated from mutually exclusive splicing of the two 3′ mini exons of the gene (Fig. 1A, B). Among 18 exons that constitute the skeletal TnT gene, three types of splicing are exhibited. Each of exons 4 to 8 near the 5′ end of the primary transcript may be individually included or excluded (depending on the species) from the mature mRNA (inclusion-exclusion splicing) in a combinatorial fashion to generate as many as 32 different sequences within the NH_2-terminal region of the protein. These protein variants are different with respect to their amino acid sequence, charge and overall size (Fig. 1B). The diversity among TnT isoforms varies within three regions of its primary sequence. Exons 2, 3, 9–15 and 18 splice constitutively, yielding amino acids 1–10, 45–236 and 251–267, respectively and are shared by all isoforms of TnT. The high degree of diversity lies in the combination of exons 4–8, coding for aa 11–44, and has been suggested to affect the interaction of TnT with Tm (Pearlstone and Smillie 1981). The segment of TnT, from aa 237–250 is encoded by either exon 16(α) or 17(β), and represents a variable domain of the protein that is thought to be involved in a Ca^{2+} sensitive interaction with TnC and TnI (Pearlstone and Smillie 1985). These two exons are developmentally regulated and β is primarily found in embryonic muscle and α in adult muscle. In rabbit skeletal muscle exons x, 6 and 7

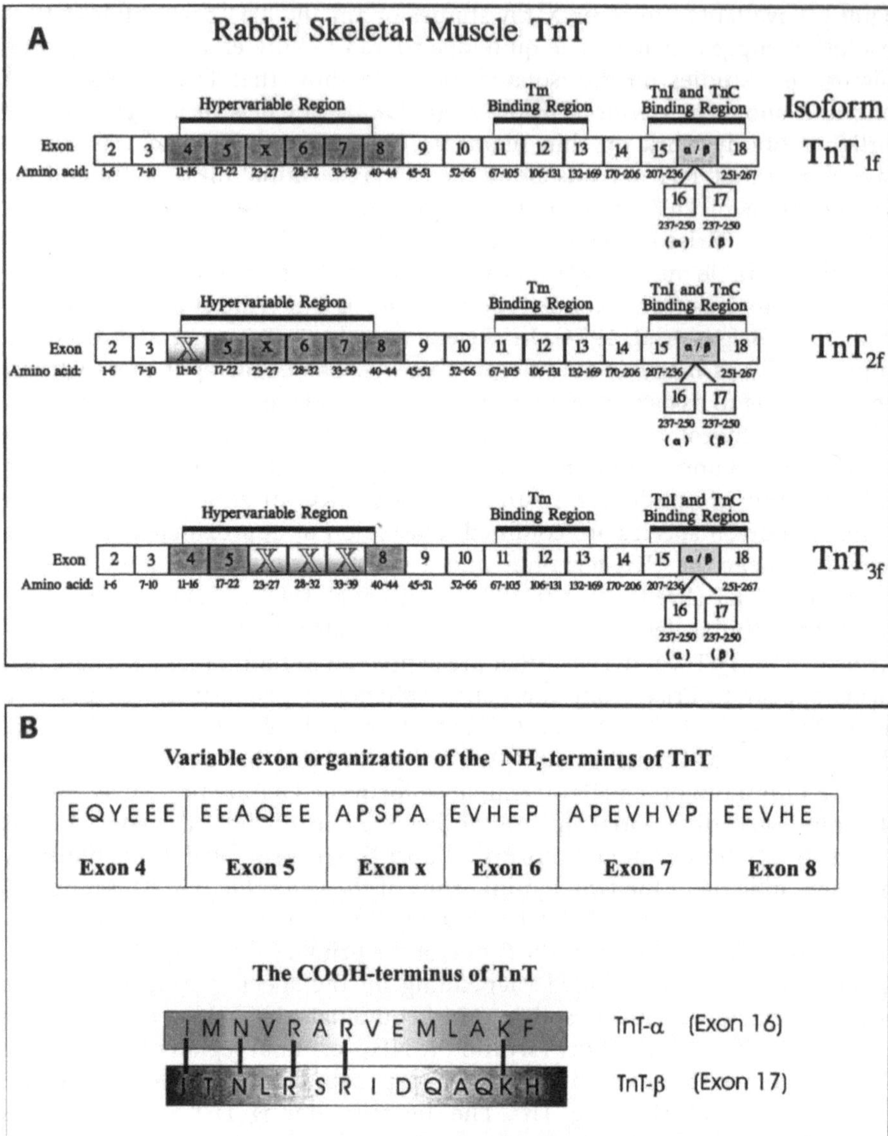

Fig. 1A,B. Exon organization of rabbit skeletal TnT. **A** NH_2-terminal isoforms (1f, 2f and 3f) of rabbit skeletal muscle TnT. *Large Xs* indicate missing exons. **B** Amino acid sequences of the NH_2- and COOH-termini of TnT

(Fig. 1A, B) splice as a unit, thereby yielding only 16 possible NH_2-terminal variants of TnT (Briggs and Schachat 1989).

Studies with proteolytic fragments of TnT have indicated that the functionally important sites of TnT are mostly located in the COOH-terminal half of the molecule. This region of TnT interacts with tropomyosin, TnC, TnI and actin (Pearlstone and Smillie 1985; Heeley et al. 1987; Ishii and Lehrer 1991). The NH_2-terminus of TnT has been shown to interact with a long stretch of the COOH-terminal region and the head-to-tail overlap region of Tm (Brisson 1986; White et al. 1987). TnT can be separated proteolytically into two fragments, TnT1 (T1) and TnT2 (T2) (Morris and Lehrer 1984), comprising of 1–158 and 159–259 (rabbit skeletal isoform) amino acid residues, respectively. TnT1 (M_r~26 kDa), an elongated part of TnT, binds to the head-to-tail overlap region of Tm while TnT2 (M_r~13 kDa) most likely has a globular shape and interacts with TnI•TnC and Tm. According to Schaertl et al. 1995, the NH_2-terminal region of TnT (T1) plays an important role in determining the size of the thin filament cooperative unit, while its COOH-terminal region (T2) together with TnI•TnC is responsible for the Ca^{2+}-sensitive binding of S1 to Tm-actin. The latter is consistent with Potter et al. 1995 for a dual role of TnT, i.e., (1) anchoring of the TnC•TnI complex to actin/Tm, and (2) activating actomyosin ATPase activity in a Ca^{2+}-dependent manner. Earlier studies of Pearlstone and Smillie (1978) and Heeley et al. (1987) have demonstrated some evidence for Ca^{2+} dependent interactions between TnC and TnT. Recent work from Stefancik et al. (1998) suggests such an interaction between TnT and TnC in the yeast two hybrid system, a genetic approach used to study protein-protein interactions. Deletions and point mutations in TnT, which disrupted a TnT heptad repeat motif, severely reduced TnT and TnI interaction, but the binding of TnT to TnC was not affected (Stefancsik et al. 1998). In another study from these authors (Jha and Sarkar 1998), residue 155 of TnT was shown to be within 10 Å of TnC in the presence or absence of Ca^{2+}.

However, nothing was known about the possible functional significance of the sequence differences between the two COOH-terminal α and β isoforms of TnT until the study of Pan and Potter 1992. Since the COOH-terminal region of TnT is known to be involved in interactions with Tm, TnI, TnC and actin, it is conceivable that the α and β isoforms of TnT may behave differently in one or more of these interactions and thereby have altered functional properties. Pan and Potter 1992, using recombinant TnT fragments corresponding to the α- and β- isoforms have characterized the differences between the α-and β-isoforms when complexed with TnC. In other studies, Pan et al. 1991, have shown the effect of the NH_2-terminal region of TnT on actin activated S1-MgATPase activity. A TnT fragment missing the first 45 amino acids demonstrated only 60% of the maximal activity of the TnT_{2f} isoform of TnT (Pan et al. 1991). This observation suggests that the NH_2-terminal region of TnT may determine the level of activation of actomyosin ATPase activity. Thus, although much is known about the structure of the various isoforms of TnT and their interactions with the other thin filament proteins, little is known about the function of TnT, nor of the possible physiological significance of its 64 potential isoforms. A recent study of Wang and Jin 1998, demonstrates that the binding of Zn^{2+} or monoclonal antibodies to the NH_2-terminal region of TnT induces extensive conformational changes in the entire molecule. The NH_2-terminal domain mediated conformational changes in TnT resulted in an altered interaction of TnT with Tm and TnI. As mentioned above, TnT complexed with TnI, TnC and Tm confers Ca^{2+} sensitivity to the actomyosin ATPase.

The role of TnT in the Ca^{2+}-mediated regulation of actin-Tm activated myosin ATPase activity is illustrated in Fig. 2 (Potter et al. 1995). It also demonstrates the role of the other troponin subunits, TnI and TnC, in this regulation. TnI alone can inhibit ATPase activity independent of calcium. This inhibition can be reversed, by the addition of TnC, back to the basal activity level, but never higher (the basal activity with F-actin and myosin is the same with or without Tm for the conditions used in this experiment). As increasing concentrations of Tn are added (Fig. 2) in the absence of Ca^{2+}, the ATPase is inhibited to the same level as with TnI alone. In contrast to the situation with TnI+TnC, addition of Tn in the presence of Ca^{2+} leads to an activation of the basal activity. Thus, this activation can only be seen in the presence of TnT in the Tn complex and in the presence of Ca^{2+}. Consistent with this, TnT+TnC (+Ca^{2+}) alone cannot activate the ATPase activity (Fig. 2), indicating a need for TnI in order to have activation. However, this activation of the actomyosin ATPase activity can occur in the absence of an interaction between TnT and TnI. The study of Sheng et al. 1992, shows that a deletion mutant of TnI, namely TnI_{d57}, that does not bind to TnT, but does interact with TnC, could still form functional complexes with TnC (in the presence of Ca^{2+}) and regulate actomyosin-S1 ATPase activity (Fig. 2). Both wild type Tn (TnI•TnC•TnT) and

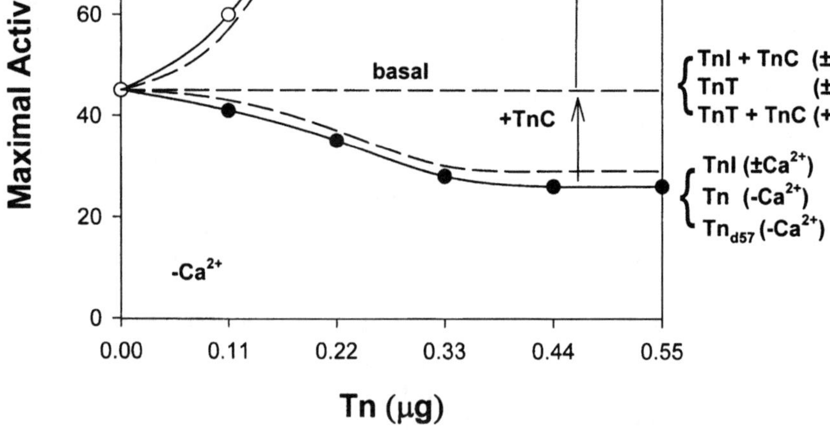

Fig. 2. The role of troponin subunits in the regulation of the actin/Tm-myosin ATPase activity

mutant Tn_{d57} (TnI_{d57}•TnC•TnT) complexes demonstrated equivalent activity in the Ca^{2+} regulation of actomyosin ATPase activity. Taken together, these results also suggest that TnT activates the ATPase activity via direct interaction with TnC, since regulation can occur in the absence of an interaction between TnI and TnT. Based on an in vitro motility approach, the Marston group comes to a similar conclusion (Bing et al. 1997). The complex of TnI•TnC inhibited actin-Tm activation of myosin ATPase activity in the absence of Ca^{2+}, but had no effect at high Ca^{2+} concentrations. Additional effects were achieved when TnT was present resulting in a 2.5-fold increase in ATPase activity at pCa5 (Bing et al. 1997).

A potential interaction site between TnC and TnT, that could be responsible for the ATPase activation, is suggested by another series of experiments (Pan and Potter 1992). Using recombinant TnT fragments (TnT2p-α and TnT2p-β) corresponding to the α and β isoforms (Fig. 1; note: large Xs indicate missing exons) of TnT (both isoforms were truncated at their NH_2-termini and contained only the last 109 amino acid residues of the COOH-terminus), the authors have shown that TnT2p-α bound to TnC more strongly than TnT2p-β in the presence and absence of Ca^{2+}. They also demonstrated that TnT2p-α had a higher affinity for Tm than TnT2p-β. The Ca^{2+} affinity of the Ca^{2+}-specific regulatory sites of TnC in a TnT2p-α•TnC complex was threefold higher than in a TnT2p-β•TnC complex. Interestingly, embryonic skeletal muscle, which contains the β isoform, is less sensitive to Ca^{2+} than α containing adult muscle (Kerrick et al. 1976). These data suggest that (1) the α/β region of TnT may be a site, or part of a site, of interaction between TnT and TnC, and/or (2) that the sequence differences between the α- and β-isoforms of TnT may contribute in part to the determination of the Ca^{2+} sensitivity of muscle contraction (Fig. 1B).

Two full-length cDNAs corresponding to the two COOH-terminal isoforms of one TnT isoform (TnT_{3f}), namely TnT_{3f}-α and TnT_{3f}-β (Fig. 1A, B), as well as a deletion mutant lacking the 13 amino acid region (no exon 16/17) which is responsible for the α/β isoform variability (TnT_{3f}-Δ), have been cloned (Panavelil et al. 1997). All of these TnTs were based on the TnT_{3f} structure (Fig. 1A). These proteins (TnT_{3f}-α, TnT_{3f}-β, and TnT_{3f}-Δ) were examined in skinned skeletal muscle fibers where the Ca^{2+} sensitivity of force development could be assessed (Panavelil et al. 1997; Potter et al. 1998). To accomplish this, the methods of Shiraishi and Yamamoto (1994) and Hatakenaka and Ohtsuki (1992) were applied where skinned fibers are treated with relatively high concentrations of TnT. This treatment results in the displacement of the endogenous Tn complex and incorporation of the added TnT. The displaced Tn activity can be reconstituted by treating the preparation containing the added TnT with a preformed TnC•TnI complex.

Figure 3A illustrates a typical experiment with a control fiber (Fig. 3B) that was run in parallel with the treated fiber and was treated with the same solutions (minus TnT and minus the TnI•TnC complex) for the same amount of time, in order to estimate the amount of rundown of the fiber during this time

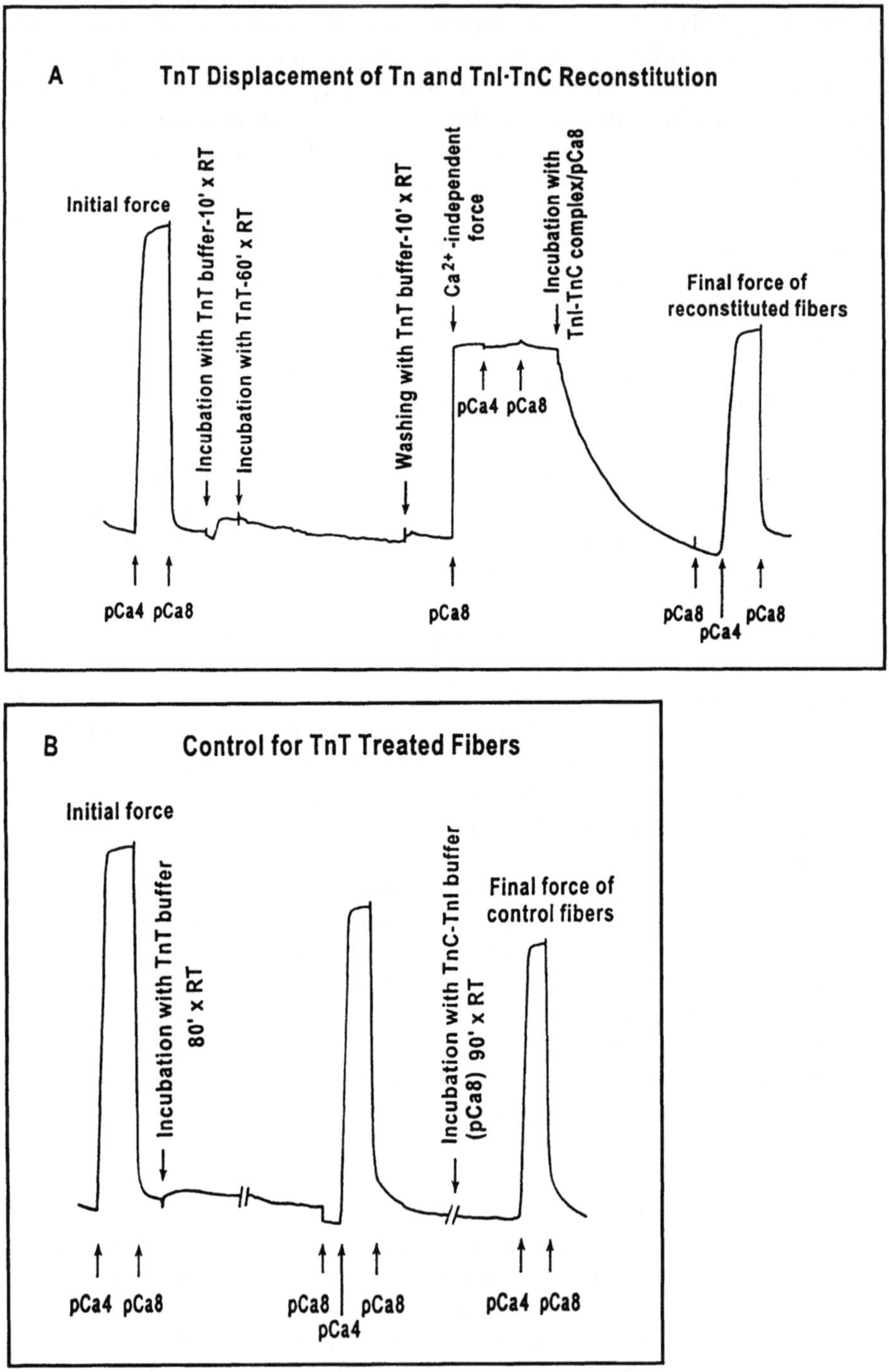

Fig. 3A,B. The protocol for the rabbit skeletal TnT (RSTnT)-treated (Tn-displaced) skinned skeletal muscle fibers. **A** Experimental, RSTnT-treated fiber. **B** Control, buffer-treated fiber

period and to serve as a baseline for calculating the percentage reconstitution of the treated fiber (Potter et al. 1998). As can be seen, treatment of the fibers with TnT resulted in a complete loss of Ca^{2+}-dependence of force and the fibers became unregulated (Fig. 3A). When these fibers were incubated with a preformed TnI•TnC complex, dissolved in the relaxing solution (pCa8), the fibers underwent a gradual relaxation as the Tn activity was reconstituted. Once fully relaxed, the force became entirely regulated by Ca^{2+}. The results support the concept that TnT plays an active role in the Ca^{2+}-regulation of contraction, and that it does more than simply anchor the TnI•TnC complex to the thin filament through TnTs interaction with Tm.

3 Model of the Regulation of Contraction by Troponin

Structural studies suggest that the binding of Ca^{2+} to the thin filament initiates the movement of tropomyosin (Tm) on actin, an observation that formed the basis for the steric blocking theory of muscle contraction (Huxley 1972; Haselgrove 1972; Parry and Squire 1973). The Ca^{2+} induced movement of Tm preceding attachments of myosin heads was observed by X-ray diffraction of frog muscles (Kress et al. 1986) or skinned non-overlapped rabbit psoas muscle fibers (Poole et al. 1994). However, the low resolution of the X-ray diffraction patterns cannot exclude a possible contribution of Ca^{2+}-induced troponin reorganization to the second layer line changes (Lehrer and Geeves 1998 and references therein). Since the atomic structure of the troponin complex is still not available, structuralists have neglected this possibility. Structural and biochemical aspects of regulation in vertebrate skeletal muscle were recently reviewed by Squire and Morris (1998). The authors also presented a modified steric blocking model for regulation where either tropomyosin or troponin or both move during activation. Interestingly, this calculation could not rule out the movement of troponin during activation (Squire and Morris 1998).

As was proposed by Schaertl et al. (1995), Lehrer and Geeves (1998), and Maytum et al. (1999), thin filaments can exist in two Ca^{2+}-sensitive equilibria: "blocked-closed" and "closed-open", and the regulation of muscle contraction involves cooperative and allosteric interactions among the contractile proteins. In the blocked state, the binding of myosin heads to actin is inhibited. Weakly bound crossbridges exist in the "closed" state but isomerization of the crossbridge to the strongly bound "R-type" is inhibited. In the "open" state, strongly bound "R-type" complexes are formed and these are required for the stimulation of ATPase activity or contraction. Troponin acts as an allosteric inhibitor in the absence of Ca^{2+} and as an allosteric activator in the presence of Ca^{2+}.

Our current hypothesis is that TnC has a dual role: (1) release of ATPase inhibition through an interaction with TnI, and (2) activation of the ATPase through an interaction with TnT. Therefore, a more detailed understanding of the functional role of skeletal TnT is of prime importance, given its potential involvement in the activation of actomyosin ATPase. Known interactions

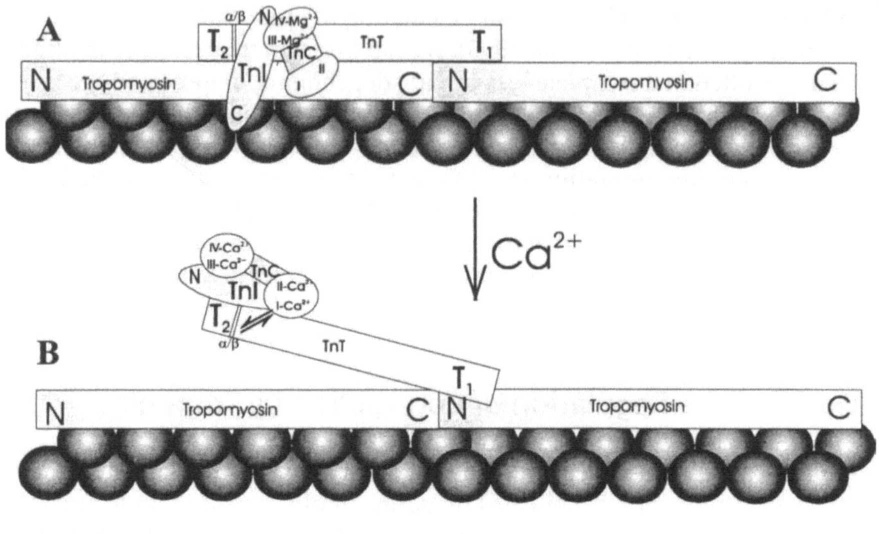

Fig. 4A,B. Troponin-tropomyosin-actin interactions in the regulation of contraction. **A** Relaxed state. **B** Ca^{2+}-activated state

between Tn, Tm, and actin could be incorporated into the model of Schaertl et al. 1995. These are schematically illustrated in Fig. 4. In the absence of Ca^{2+} (Fig. 4A), TnC interacts with TnI in an antiparallel fashion. Mg^{2+} bound to the Ca^{2+}-Mg^{2+} sites of TnC maintains the interaction between the COOH-terminus of TnC and the NH_2-terminus of TnI. The COOH-terminus of TnI is bound to the actin-Tm complex, holding it in the "blocked state". TnT has two potential binding sites on Tm (Fig. 4A). One is between the NH_2-terminus (T1) of TnT and the overlap region between the COOH- and NH_2-termini of two adjacent Tm molecules. The second is between the COOH-terminus (T2) of TnT and the central region of Tm. Whereas the former is thought to be constant, the latter is thought to depend on an interaction between TnC and T2, i.e., in the absence of Ca^{2+}, T2 is bound to Tm and acts as an inhibitor (Fig. 4A); in the presence of Ca^{2+}, TnC binds to T2 and causes it to dissociate from Tm, removing the inhibitory activity and allowing activation (Fig. 4B). Therefore, Ca^{2+} binding to the Ca^{2+}-specific sites of TnC would generate two new interactions for TnC (Fig. 4B). The first is an interaction between the NH_2-terminus of TnC and the COOH-terminus of TnI (including the inhibitory region) which causes the dissociation of TnI from the actin-Tm complex and leads to the "closed" state of the thin filament. This would allow weak interactions with myosin heads and basal actin activated myosin ATPase activity (Fig. 2). The second interaction is between the NH_2-terminus of TnC and the COOH-terminus (T2) of TnT which weakens the interaction between T2 and Tm and leads to an activation (through dis-inhibition) of the ATPase activity (Fig. 2), and would be equivalent to the "open" state of Geeves and Conibear (1995), and Schaertl et al. (1995).

Residues 1–191 of TnT containing the T1 and a part of T2 regions were identified by Malnic et al. (1998), to posses an intrinsic activation activity. A recombinant TnT fragment, TnT_{1-191}, was able to activate actomyosin ATPase activity on its own without the TnI•TnC complex. The authors also suggest an inhibitory domain of TnT being located somewhere between residues 191–263. The full-length TnT could activate the ATPase activity only in the presence of Ca^{2+} and TnI•TnC complex. This would be expected due to the intrinsic activation/inhibitory properties of TnT and is totally consistent with the original proposal of Potter et al. (1995). Two classes of interactions between Tn subunits have been proposed: the structural, Ca^{2+}-independent interactions occurring between the COOH-terminal domain of TnC, the NH_2-terminal domain of TnI and the last 50 (216–263) residues of TnT and the regulatory, Ca^{2+}-dependent interactions between the NH_2-terminal, Ca^{2+}-specific domain of TnC, the COOH-terminal domain of TnI (containing the inhibitory region) and the region of TnT spanning the residues 157–216 (Malnic et al. 1998). These findings are in agreement with Jha et al. 1996, whose COOH-terminal deletion mutant of TnT (TnT_{1-201}) demonstrated reduced complex formation with TnC and TnI and decreased Ca^{2+} sensitivity in the reconstituted regulatory system.

The dependence of the Ca^{2+}-sensitizing function of TnT on the amino acid sequence of its NH_2-terminus remains an open question (Perry 1998). The large number of TnT isoforms compared with other thin filament proteins suggests the importance of the NH_2-terminal variable region of TnT in the Ca^{2+}-regulation of skeletal muscle contraction. It is anticipated that we will learn more about the functional and physiological significance of the multiple TnT isoforms over the coming years. It is also likely that detailed structural information will be forthcoming once the crystal structures of intact complexes of TnC, TnI and TnT are known.

References

Abe H, Komiya T, Obinata T (1986) Expression of multiple troponin T variants in neonatal chicken breast muscle. Dev Biol 118:42–51

Babu A, Rao VG, Su H, Gulati J (1993) Critical minimum length of the central helix in troponin C for the Ca^{2+} switch in muscular contraction. J Biol Chem 268:19232–19238

Bing W, Fraser ID, Marston SB (1997) Troponin I and troponin T interact with troponin C to produce different Ca^{2+}-dependent effects on actin-tropomyosin filament motility. Biochem J 327:335–340

Breitbart RE, Nguyen HT, Medford RM, Destree AT, Mahdavi V, Nadal-Ginard B (1985) Intricate combinatorial patterns of exon splicing generate multiple regulated troponin T isoforms from a single gene. Cell 41:67–82

Briggs MM, Schachat F (1989) N-terminal amino acid sequences of three functionally different troponin T isoforms from rabbit fast skeletal muscle. J Mol Biol 206:245–249

Briggs MM, Maready M, Schmidt JM, Schachat F (1994) Identification of a fetal exon in the human fast troponin T gene. FEBS Lett 350:37–40

Brisson JR (1986) Interaction of tropomyosin and troponin T: a proton nuclear magnetic resonance study. Biochemistry 25:4548–4555

Campbell AP, Sykes BD (1991) Interaction of troponin I and troponin C. Use of the two-dimensional nuclear magnetic resonance transferred nuclear Overhauser effect to determine the structure of the inhibitory troponin I peptide when bound to skeletal troponin C. J Mol Biol 222:405–421

Chandra M, da Silva EF, Sorenson MM, Ferro JA, Pearlstone JR, Nash BE, Borgford T, Kay CM, Smillie LB (1994) The effects of N helix deletion and mutant F29W on the Ca^{2+} binding and functional properties of chicken skeletal muscle troponin C. J Biol Chem 269:14988–14994

Chandra M, McCubbin WD, Oikawa K, Kay CM, Smillie LB (1994) Ca^{2+}, Mg^{2+}, and troponin I inhibitory peptide binding to a Phe-154 to Trp mutant of chicken skeletal muscle troponin C. Biochemistry 33:2961–2969

Cooper TA, Ordahl CP (1985) A single cardiac troponin T gene generates embryonic and adult isoforms via developmentally regulated alternate splicing. J Biol Chem 260:11140–11148

Dalgarno D, Grand RJA, Levine BA, Moir AJG, Scott GMM, Perry SV (1982) Interaction between troponin I and troponin C. Definition of the topography by proton magnetic resonance spectroscopy. FEBS Lett 150:54–58

Dobrowolski Z, Xu G-Q, Hitchcock-DeGregori SE (1991) Modified calcium-dependent regulatory function of troponin C central helix mutants. J Biol Chem 266:5703–5710

Farah CS, Miyamoto CA, Ramos CHI, da Silva ACR, Quaggio RB, Fujimori K, Smillie LB, Reinach FC (1994) Structural and regulatory functions of the NH_2- and COOH-terminal regions of skeletal muscle troponin I. J Biol Chem 269:5230–5240

Flicker PF, Phillips GN Jr, Cohen C (1982) Troponin and its interactions with tropomyosin. An electron microscope study. J Mol Biol 162:485–501

Francois JM, Sheng Z, Szczesna D, Potter JD (1995) The functional role of the domains of troponin-C investigated with thrombin fragments of troponin-C reconstituted into skinned muscle fibers. J Biol Chem 270:19287–19293

Gagne SM, Tsuda S, Li MX, Chandra M, Smillie LB, Sykes BD (1994) Qualification of the calcium-induced secondary structural changes in the regulatory domain of troponin-C. Protein Sci 3:1961–1974

Gagne SM, Tsuda S, Li MX, Smillie LB, Sykes BD (1995) Structures of troponin C regulatory domains in the apo and calcium-saturated states. Nature Struct Biol 2:784–789

Gagne SM, Li MX, Sykes BD (1997) Mechanism of direct coupling between binding and induced structural change in regulatory calcium binding proteins. Biochemistry 36:4386–4392

Gagne SM, Tsuda S, Spyracopoulos L, Kay LE, Sykes BD (1998) Backbone and methyl dynamics of the regulatory domain of troponin C: anisotropic rotational diffusion and contribution of conformational entropy to calcium affinity. J Mol Biol 278:667–686

Geeves MA, Conibear PB (1995) The role of three-state docking of myosin S1 with actin in force generation. Biophys J 68:194s–201s

Gergely J (1998) Molecular switches in troponin. Adv Exp Med Biol 453:169–176

Grabarek Z, Leavis PC, Gergely J (1986) Calcium binding to the low affinity sites in troponin C induces conformational changes in the high affinity domain: a possible route of information transfer in activation of muscle contraction. J Biol Chem 261:608–613

Grabarek Z, Tao T, Gergely J (1992) Molecular mechanism of troponin-C function. J Muscle Res Cell Motil 13:383–393

Greaser ML, Gergely J (1971) Reconstitution of troponin activity from three protein components. J Biol Chem 246:4226–4233

Gulati J, Babu A, Su H, Zhang Y-F (1993) Identification of the regions conferring calmodulin-like properties to troponin C. J Biol Chem 268:11685–11690

Haselgrove JC (1972) X-ray evidence for a conformational change in actin-containing filaments of vertebrate striated muscle. Cold Spring Harbor Symp Quant Biol 37:341–352

Hatakenaka M, Ohtsuki I (1992) Effect of removal and reconstitution of troponins C and I on the Ca^{2+}-activated tension development of single glycerinated rabbit skeletal muscle fiber. Eur J Biochem 205:985–993

Heeley DH, Golosinska K, Smillie LB (1987) The effects of troponin T fragments T1 and T2 on the binding of nonpolymerizable tropomyosin F-actin in the presence and absence of troponin I and troponin C. J Biol Chem 262:9971–9978
Hernandez G, Blumenthal DK, Kennedy MA, Unkefer CJ, Trewhella J (1999) Troponin I inhibitory peptide (96–115) has an extended conformation when bound to skeletal muscle troponin C. Biochemistry 38:6911–6917
Herzberg O, James MNG (1985) Structure of the calcium regulatory muscle protein-troponin C at 2.8 Å resolution. Nature 313:653–659
Herzberg O, James MNG (1988) Refined crystal structure of troponin C from turkey skeletal muscle at 2.0-Å resolution. J Mol Biol 203:761–769
Herzberg O, Moult J, James MNG (1986) A model for the Ca^{2+}-induced conformational transitions of troponin C. J Biol Chem 261:2638–2644
Houdusse A, Love ML, Dominguez R, Grabarek Z, Cohen C (1997) Structures of four Ca^{2+}-bound troponin C at 2.0-Å resolution: further insights into Ca^{2+}-switch in the calmodulin superfamily. Structure 5:1695–1711
Huxley HE (1972) Structural changes in the actin- and myosin-containing filaments during contraction. Cold Spring Harbor Symp Quant Biol 37:361–376
Ishii Y, Lehrer SS (1991) Two-site attachment of troponin to pyrene-labeled tropomyosin. J Biol Chem 266:6894–6903
Jha PK, Sarkar S (1998) A recombinant monocysteine mutant (Ser to Cys-155) of fast skeletal troponin T: identification by cross-linking of a domain involved in a physiologically relevant interaction with troponins C and I. Biochemistry 37:12253–12260
Jha PK, Leavis PC, Sarkar S (1996) Interaction of deletion mutants of troponins I and T: COOH-terminal truncation of troponin T abolishes troponin I binding and reduces Ca^{2+} sensitivity of the reconstituted regulatory system. Biochemistry 35:16573–16580
Johnson JD, Collins JH, Potter JD (1978) Dansylaziridine-labeled troponin C: a fluorescent probe of Ca^{2+}-specific regulatory sites. J Biol Chem 253:3775–3777
Kerrick WGL, Secrist D, Coby R, Lucas S (1976) Development of difference between red and white muscles in sensitivity to Ca^{2+} in the rabbit from embryo to adult. Nature 260:440–441
Kobayashi T Grabarek Z, Gergely J, Collins JH (1995) Extensive interactions between troponins C and I. Zero-length cross-linking of troponin I and acetylated troponin C. Biochemistry 34:10946–10952
Kress M, Huxley HE, Faruqi AR, Hendrix J (1986) Structural changes during activation of frog muscle studies by time-resolved X-ray diffraction. J Mol Biol 188:325–342
Lehrer SS, Geeves M (1998) The muscle thin filament as a classical cooperative/allosteric regulatory system. J Mol Biol 277:1081–1089
Leszyk J, Collins JH, Leavis PC, Tao T (1987) Cross-linking of rabbit skeletal muscle troponin with the photoactive reagent 4-maleimidobenzophenone: identification of residues in troponin I that are close to cysteine-98 of troponin C. Biochemistry 26:7042–7047
Leszyk J, Grabarek Z, Gergely J, Collins J (1990) Characterization of zero-length cross-links between rabbit skeletal muscle troponin C and troponin I: evidence for a direct interaction between the inhibitory region of troponin I and the NH_2-terminal, regulatory domains of troponin C. Biochemistry 29:299–304
Leszyk J, Tao T, Nuwaysir L, Gergely J (1998) Identification of the photo-cross-linking sites in troponin I with 4-maleidobenzophenone labelled mutant troponin-Cs having single cysteines at positions 158 and 21. J Muscle Res Cell Motil 19:479–490
Liu W, Dotson DG, Lin X, Mullen III JJ, Gonzalez-Gray ML, Lu Q, Putkey JA (1994) The presence but not the sequence of the N-terminal peptide in cardiac TnC is important for function. FEBS Lett 347:152–156
Luo Y, Leszyk J, Qian Y, Gergely J, Tao T (1999) Residues 48 and 82 at the N-terminal hydrophobic pocket of rabbit skeletal muscle troponin-C photo-cross-link to Met121 of troponin-I. Biochemistry 38:6678–6688
Malnic B, Farah CS, Reinach FC (1998) Regulatory properties of the NH_2- and COOH-terminal domains of troponin T. J Biol Chem 273:10594–10601

Maytum R, Lehrer SS, Geeves MA (1999) Cooperativity and switching within the three-state model of muscle regulation. Biochemistry 38:1102–1110

McKay RT, Tripet BP, Hodges RS, Sykes BD (1997) Interaction of the second binding region of troponin I with the regulatory domain of skeletal muscle troponin C as determined by NMR spectroscopy. J Biol Chem 272:28494–28500

McKay RT, Pearlstone JR, Corson DC, Gagne SM, Smillie LB, Sykes BD (1998) Structure and interaction site of the regulatory domain of troponin C when complexed with 96–148 region of troponin-I. Biochemistry 37:12419–12430

Medford RM, Nguyen HT, Destree AT, Summers E, Nadal-Ginard B (1984) A novel mechanism of alternative RNA splicing for the developmentally regulated generation of troponin T isoforms from a single gene. Cell 38:409–421

Moncrieffe M, Eaton S, Bajzer Z, Haydock C, Potter JD, Laue TM, Prendergast FG (1999a) Rotational and translational motion of troponin C. J Biol Chem 274:17464–17470

Moncrieffe M, Venyaminov SY, Miller T, Guzman G, Potter JD, Prendergast FG (1999b) Optical spectroscopic characterization of single tryptophan mutants of chicken skeletal troponin C: evidence for inter-domain interaction. Biochemistry 38:11973–11983

Morris EP, Lehrer SS (1984) Troponin–tropomyosin interactions. Fluorescence studies of the binding of troponin, troponin T, and chymotryptic troponin T fragments to specifically labeled tropomyosin. Biochemistry 23:2214–2220

Moss RL (1992) Ca^{2+} regulation of mechanical properties of striated muscle. Mechanistic studies using extraction and replacement of regulatory proteins. Circ Res 70:865–884

Ohtsuki I (1979) Molecular arrangement of troponin-T in the thin filament. J Biochem 86:491–497

Ohtsuki I (1999) Calcium ion regulation of muscle contraction: the regulatory role of troponin T. Mol Cell Biochem 190:33–38

Ohtsuki I, Maruyama K, Ebashi S (1986) Regulatory and cytoskeletal proteins of vertebrate skeletal muscle. Adv Protein Chem 38:1–67

Olah GA, Rokop SE, Wang C-LA, Blechner SL, Trewhella J (1994) Troponin I encompasses an extended troponin C in the Ca^{2+}-bound complex: a small-angle X-ray and neutron scattering study. Biochemistry 33:8233–8239

Pan B-S, Potter JD (1992) Two genetically expressed troponin T fragments representing the α and β isoforms exhibit functional differences. J Biol Chem 267:23052–23056

Pan B-S, Gordon AM, Potter JD (1991) Deletion of the first 45 NH_2-terminal residues of rabbit skeletal troponin T strengthens binding of troponin to immobilized tropomyosin. J Biol Chem 266:12432–12438

Panavelil T, Guzman G, Jones M, Pan, B-S, Szczesna D, Potter JD (1997) Structural elements of the COOH-terminal region of troponin-T involved in the regulation of skeletal muscle contraction. Biophys J 72:A60

Parry DAD, Squire JM (1973) Structural role of tropomyosin in muscle regulation: analysis of the X-ray diffraction patterns from relaxed and contracting muscles. J Mol Biol 75:33–55

Pearlstone JR, Smillie LB (1978) Troponin T fragments: physical properties and binding to troponin C. Can J Biochem 56:521–527

Pearlstone JR, Smillie LB (1981) Identification of a second binding region on rabbit skeletal troponin-T for alpha-tropomyosin. FEBS Lett 128:119–122

Pearlstone JR, Smillie LB (1985) The interaction of rabbit skeletal muscle troponin-T fragments with troponin-I. Can J Biochem Cell Biol 63:212–218

Pearlstone JR, Sykes BD, Smillie LB (1997) Interactions of structural C and regulatory N domains of troponin C with repeated sequence motifs in troponin I. Biochemistry 36:7601–7606

Perry SV (1998) Troponin T: genetics, properties and function. J Muscle Res Cell Motil 19:575–602

Poole KJV, Lorenz M, Evans G, Rosenbaum G, Holmes KC (1994) The effect of calcium on the regulated thin filament structure. Biophys J 66:A347

Potter JD, Gergely J (1975) The calcium and magnesium binding sites on troponin and their role in the regulation of myofibrillar ATPase. J Biol Chem 250:4628–4633

Potter JD, Sheng Z, Pan B-S, Zhao J (1995) A direct regulatory role for troponin T and a dual role for troponin C in the Ca^{2+} regulation of muscle contraction. J Biol Chem 270:2557–2562

Potter JD, Panavelil T, Guzman G, Jones M, Zhao J, Szczesna D (1998) The role of troponin T in the regulation of contraction. Biophys J 74:A143

Ramakrishnan S, Hitchcock-DeGregori SE (1995) Investigation of the structural requirements of the troponin C central helix for function. Biochemistry 34:16789–16796

Ramakrishnan S, Hitchcock-DeGregori SE (1996) Structural and functional significance of aspartic acid 89 of the troponin C central helix in Ca^{2+} signaling. Biochemistry 35:15515–15525

Ramos CH (1999) Mapping subdomains in the C-terminal region of troponin I involved in its binding to troponin C and to thin filament. J Biol Chem 274:18189–18195

Robertson SP, Johnson JD, Potter JD (1981) The time-course of Ca^{2+} exchange with calmodulin, troponin, parvalbumin and myosin in response to transient increases in Ca^{2+}. Biophys J 34: 559–569

Schaertl S, Lehrer SS, Geeves MA (1995) Separation and characterization of the two functional regions of troponin involved in muscle thin filament regulation. Biochemistry 34:15890–15894

She M, Xing J, Dong WJ, Umeda PK, Cheung HC (1998) Calcium binding to the regulatory domain of skeletal muscle troponin C induces a highly constrained open conformation. J Mol Biol 281:445–452

Sheng Z, Strauss W, Francois JM, Potter JD (1990) Evidence that both Ca^{2+} specific sites of skeletal muscle TnC are required for full activity. J Biol Chem 265:21554–21559

Sheng Z, Francois JM, Hitchcock-DeGregori SE, Potter JD (1991) Effects of mutations in the central helix of troponin C on its biological activity. J Biol Chem 266:5711–5715

Sheng Z, Pan B-S, Miller T, Potter JD (1992) Isolation expression and mutation of a rabbit skeletal muscle cDNA clone for troponin I. The role of the NH_2-terminus of fast skeletal muscle TnI in its biological activity. J Biol Chem 267:25407–25413

Shiraishi F, Yamamoto K (1994) The effect of partial removal of troponin I and C on the Ca^{2+}-sensitive ATPase activity of rabbit skeletal myofibrils. J Biochem 115:171–173

Smillie LB, Golosinska K, Reinach FC (1988) Sequences of complete cDNAs encoding four variants of chicken skeletal muscle troponin T. J Biol Chem 263:18816–18820

Smith L, Greenfield NJ, Hitchcock-DeGregori SE (1994) The effects of deletion of the amino-terminal helix on troponin C function and stability. J Biol Chem 269:9857–9863

Smith L, Greenfield NJ, Hitchcock-DeGregori SE (1999) Mutations in the N- and D-helices of the N-domain of troponin C affect the C-domain and regulatory function. Biophys J 76:400–408

Squire JM, Morris EP (1998) A new look at thin filament regulation in vertebrate skeletal muscle. FASEB J 12:761–771

Stefancsik R, Jha PK, Sarkar S (1998) Identification and mutagenesis of a highly conserved domain in troponin T responsible for troponin I binding: potential role for coiled coil interaction. Proc Natl Acad Sci USA 95:957–962

Stone DB, Timmins PA, Schneider DK, Krylova I, Ramos CHI, Reinach FC, Mendelson RA (1998) The effect of regulatory Ca^{2+} on the in situ structures of troponin C and troponin I: a neutron scattering study. J Mol Biol 281:689–704

Strynadka NC, Chernaia M, Sielecki AR, Li MX, Smillie LB, James MNG (1997) Structural details of a calcium-induced molecular switch: X-ray crystallographic analysis of the calcium-saturated N-terminal domain of troponin C at 1.75-Å resolution. J Mol Biol 273: 238–255

Syska H, Wilkinson JM, Grand RJA, Perry SV (1976) The relationship between biological activity and primary structure of troponin I from white skeletal muscle of the rabbit. Biochem J 153:375–387

Szczesna D, Guzman G, Miller T, Zhao J, Farokhi, K, Ellemberger H, Potter JD (1996) The role of the four Ca^{2+} binding sites of troponin C in the regulation of skeletal muscle contraction. J Biol Chem 271:8381–8836

Szczesna D, Zhang R, Zhao J, Jones M, Potter JD (1999) The role of the N- and C-terminal domains of the inhibitory region of TnI in the regulation of contraction. J Biol Chem 274:29536–29542

Talbot JA, Hodges RS (1979) Synthesis and biological activity of an icosapeptide analog of the actomyosin ATPase inhibitory region of troponin I. J Biol Chem 254:3720–3723

Tripet B, Van Eyk JE, Hodges RS (1997) Mapping of a second actin-tropomyosin and second troponin C binding site within the C terminus of troponin I, and their importance in the Ca^{2+}-dependent regulation of muscle contraction. J Mol Biol 271:728–750

Van Eyk JE, Kay CM, Hodges RS (1991) A comparative study of the interactions of synthetic peptides of the skeletal and cardiac troponin I inhibitory region with skeletal and cardiac troponin C. Biochemistry 30:9974–9981

Van Eyk JE, Thomas LT, Tripet B, Wiesner RJ, Pearlstone JR, Farah CS, Reinach FC, Hodges RS (1997) Distinct regions of troponin I regulate Ca^{2+}-dependent activation and Ca^{2+} sensitivity of the acto-S1-Tm ATPase activity of the thin filament. J Biol Chem 272:10529–10537

Vassylyev DG, Takeda S, Wakatsuki S, Maeda K, Maeda Y (1998) Crystal structure of troponin C in complex with troponin I fragment at 2.3-Å resolution. Proc Natl Acad Sci USA 95:4847–4852

Wang J, Jin J-P (1998) Conformational modulation of troponin T by configuration of the NH_2-terminal variable region and functional effects. Biochemistry 37:14519–14528

Wang CK, Liao R, Cheung HC (1993) Rotational dynamics of skeletal muscle troponin C. J Biol Chem 268:14671–14677

White SP, Cohen C, Phillips JN Jr (1987) Structure of co-crystals of tropomyosin and troponin. Nature 325:826–828

Zot HG, Potter JD (1982) A structural role for the Ca^{2+}-Mg^{2+} sites of Troponin C (TnC) in the regulation of muscle contraction: preparation and properties of TnC-depleted myofibrils. J Biol Chem 257:7678–7683

Zot AS, Potter JD (1987) Structural aspects of troponin-tropomyosin regulation of skeletal muscle contraction. Annu Rev Biophys Chem 16:535–559

Structural Changes Between Regulatory Proteins and Actin: A Regulation Model by Tropomyosin-Troponin Based on FRET Measurements

Masao Miki[1]

1 Introduction

Fluorescence resonance energy transfer (FRET) measurements have been used to determine the spatial relationships between regulatory proteins and actin. In this chapter, we review our recent FRET measurements (Miki et al. 1998a,b) and other related papers, and then introduce a model for the Ca^{2+}-mediated regulation of tropomyosin-troponin in skeletal muscle.

1.1 History of Ca^{2+} Regulation

Ebashi and his colleagues (1969) showed that skeletal muscle contraction is regulated by tropomyosin and troponin on actin filaments in response to a change in Ca^{2+} concentration from approximately 10^{-7} to 10^{-5} M. Troponin serves as a molecular switch in the regulation of skeletal muscle contraction. Troponin is a complex of three proteins: TnC, which binds Ca^{2+}; TnI, which inhibits actomyosin ATPase activity; and TnT, which binds tropomyosin. The binding of Ca^{2+} to TnC induces a conformational change in troponin, which is propagated along the actin filament by tropomyosin. Numerous studies have characterized the interaction between the thin filament proteins to deduce how the Ca^{2+}-triggering signal is propagated from TnC to the rest of the thin filament (for reviews, see Zot and Potter 1987; Farah and Reinach 1995). Nevertheless, the detailed molecular mechanism of this regulatory process is still not well understood.

X-ray diffraction studies of muscle fibers showed a significant change in the layer line intensities depending on whether the muscle was in a relaxed or contracting state (Haselgrove 1972; Huxley 1972). This intensity change has been interpreted in terms of an azimuthal shift of the tropomyosin strands in the grooves of the actin helix. Based on this interpretation, the steric blocking theory has been proposed, in which tropomyosin sterically blocks the myosin-binding site on the thin filament during relaxation. Although widely accepted, there is no categorical proof for the steric blocking model. The known atomic structure of actin (Kabsch et al. 1990) has been fitted to low-angle X-ray

[1] Department of Applied Chemistry and Biotechnology, Fukui University, 3-9-1 Bunkyo, Fukui-Shi 910-8507 Japan

Results and Problems in Cell Differentiation, Vol. 36
D.D. Thomas, C.G. dos Remedios (Eds.): Molecular Interactions of Actin

diffraction data (Holmes 1995) and three-dimensional reconstructions from electron microscopy (Lehman et al. 1995). Taken together, these observations strongly suggest that tropomyosin movement on actin filaments is an important component of Ca^{2+} regulation.

1.2 FRET as an Optical Ruler

FRET has been extensively used for studying the spatial relationships between residues on muscle proteins (see reviews: dos Remedios et al. 1987; Miki et al. 1992; dos Remedios and Moens 1995a). This method is especially valuable for detecting small conformational changes, since the transfer efficiency is a function of the inverse of the sixth power of the distance between probes. Using this method, several attempts have been made to detect conformational changes of thin filaments in response to a change in Ca^{2+} concentration. The distance between probes on Cys-133 of TnI and Cys-374 or Lys-61 on actin were measured in the presence and absence of Ca^{2+} (Miki 1990; Tao et al. 1990; Miki and Iio 1993) and significant differences in the distances were detected in response to a change in Ca^{2+} concentration. The notion of tropomyosin movement on the reconstituted thin filament as proposed in the steric blocking theory has also been tested using FRET (Miki and Mihashi 1979; Tao et al. 1983; Miki 1990) but no change in the distance between probes on tropomyosin and actin has ever been detected in response to a change in Ca^{2+} concentration.

1.3 Probe Sites on Proteins

The α-chain of tropomyosin contains a single cysteine (Cys-190) that can be modified without greatly altering its biological activity, depending on the reagent. Until now, Cys-190 has served as the major structural probe site for tropomyosin. In order to get more information about the structural changes between regulatory proteins and actin, new sites for FRET measurements were engineered into troponin and tropomyosin by expressing mutant isoforms. A mutant of rabbit skeletal muscle TnI was expressed in which the natural cysteines (48, 64, and 133) were replaced with Ala or Ser and a single Cys residue was introduced at position 9 near the N-terminus (Zhao et al. 1995). A mutant of rabbit skeletal muscle α-α-tropomyosin was expressed in *Escherichia coli* in which two residues were replaced (Ser-87-Cys; Cys-190-Ile) and three additional residues Met-Ala-Ser- were inserted at the N-terminus in order to restore its binding ability to F-actin (Monteiro et al. 1994). The mutant TnI (Cys-9) and tropomyosin (Cys-87), in addition to the sites of the wild-type TnI (Cys-133), TnC (Cys-98), and tropomyosin (Cys-190) were labeled with an energy donor probe, 5-(2-iodoacetylaminoethyl)-aminonaphthalene-1-sulfonic acid (IAEDANS). These sites are indicated on a schematic model of the tropomyosin and troponin complex in Fig. 1.

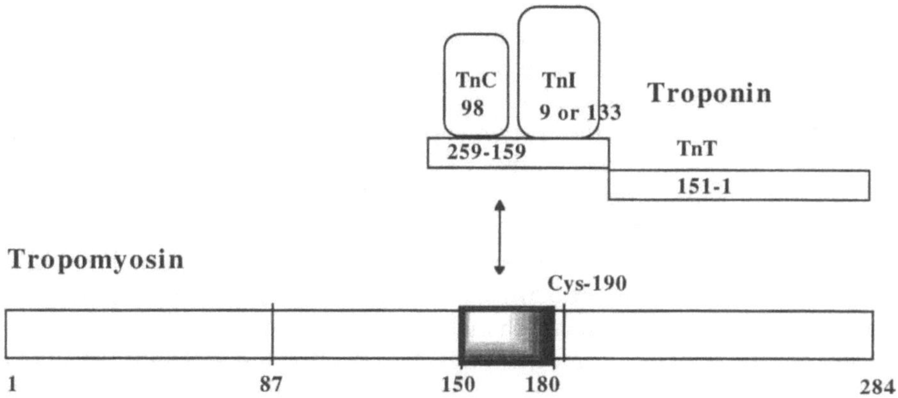

Fig. 1. Locations of the energy donor-labeling sites on a schematic model of the tropomyosin and troponin complex. The globular part of troponin composed of TnC and TnI binds to tropomyosin near Cys-190 (150–180) and the elongated part composed of TnT covers nearly one third of the tropomyosin. (See review Zot and Potter 1987)

Actin was labeled with the acceptor at Gln-41 using fluorescein cadaverine (FLC), Lys-61 using fluorescein 5-isothiocyanate (FITC), Cys-374 using 4-dimethylaminophenylazophenyl 4′-maleimide (DABMI), and the nucleotide-binding site using 2′(or 3′)-*O*-(2,4,6-trinitrophenyl)adenosine 5′-diphosphate (TNP-ADP). The ATPase activities of reconstituted systems composed of donor-labeled troponin or tropomyosin and acceptor-labeled actin were measured in the presence and absence of Ca^{2+}. In all cases, strong Ca^{2+} sensitivity of the ATPase activity was exhibited, indicating that the donor-labeled troponin or tropomyosin and acceptor-labeled actin retain their essential properties, being able to participate in the calcium regulation process. The FRET efficiency between these probes was high, and therefore they should be very sensitive to a change in the distance between the troponin- or tropomyosin-bound donor and the actin-bound acceptor. Structural changes between regulatory proteins and actin on a reconstituted thin filament was measured by FRET under physiological conditions in the presence (+Ca state; 50 μM $CaCl_2$) and absence (–Ca state; 1 mM EGTA) of Ca^{2+} ions. The energy transfer efficiency was determined by measuring the donor-fluorescence quenching in the presence of acceptor.

2 Structural Changes Measured by FRET

2.1 Tropomyosin–Actin

The steric blocking mechanism clearly demands a movement of tropomyosin relating to actin and, in principle, this should have been readily detected by probes on these two proteins. However, the distance between Cys-190 of tropomyosin and Cys-374 or the nucleotide-binding site of actin in the reconstituted thin filament did not appear to change appreciably upon the binding of Ca^{2+} to troponin (Miki and Mihashi 1979; Tao et al. 1983). Tao et al. (1983) suggested a case where the tropomyosin-bound donor and the actin-bound acceptors can be located relative to each other so that the movement of tropomyosin with respect to the grooves of actin produces no net change in the extent of energy transfer. But such acceptor locations on the actin filament are restricted to a narrow area. Later, the distance between Cys-190 of tropomyosin and Lys-61 of actin was measured in the presence and absence of Ca^{2+} ions, and only a small change (3 Å) was observed (Miki 1990). Since Lys-61 is located at 39 Å away from Cys-374 on actin, the possibility of a large-scale movement of tropomyosin not producing significant change in the FRET efficiency should be very small.

Cys-190 of tropomyosin is located close to the troponin-binding site, so the possibility still remains that the labeled region may not be very sensitive to movement of tropomyosin on the thin filament. The new site (Cys-87) of mutant tropomyosin is located far from Cys-190 and also from the troponin-binding site (see Fig. 1). Figure 2 shows two monomers of F-actin from the Lorenz et al. (1993) model, in which Cα of several amino acid residues are indicated by filled circles. The residues, which can be labeled with the FRET acceptor probes, are dispersed on the outer domain of actin in this thin filament. The FRET efficiencies and distances between these probes are summarized in Table 1.

Most efficiencies were close to 50% where FRET is most sensitive to changes in distance between probes. No significant change in the distances was observed between the presence and absence of Ca^{2+}. The distance between Lys-61 of actin and Cys-87 or Cys-190 of tropomyosin was slightly dependent on

Table 1. Distances between probes attached to tropomyosin and actin in the reconstituted thin filament in the presence and absence of Ca^{2+}

Donor site	Acceptor site (actin)	R_0 (2/3) (Å)	Efficiency (+Ca/−Ca)	R(2/3) (Å) (+Ca/−Ca)
Cys-87 of mutantTm	Nucleotide[a]	41.4	0.72	35.2
	Gln-41[a]	47.2	0.63	43.2
	Lys-61[a]	47.1	0.42/0.49	49.7/47.4
	Cys-374[a]	40.0	0.32	45.4
Cys-190 of Tm	Nucleotide[a]	38.0	0.32	43.1
	Gln-41[a]	43.0	0.25	51.6
	Lys-61[b]	45	0.46/0.55	46/43
	Cys-374[b]	46	0.35	51

[a] Data from Miki et al. (1998b).
[b] Data from Miki (1990).

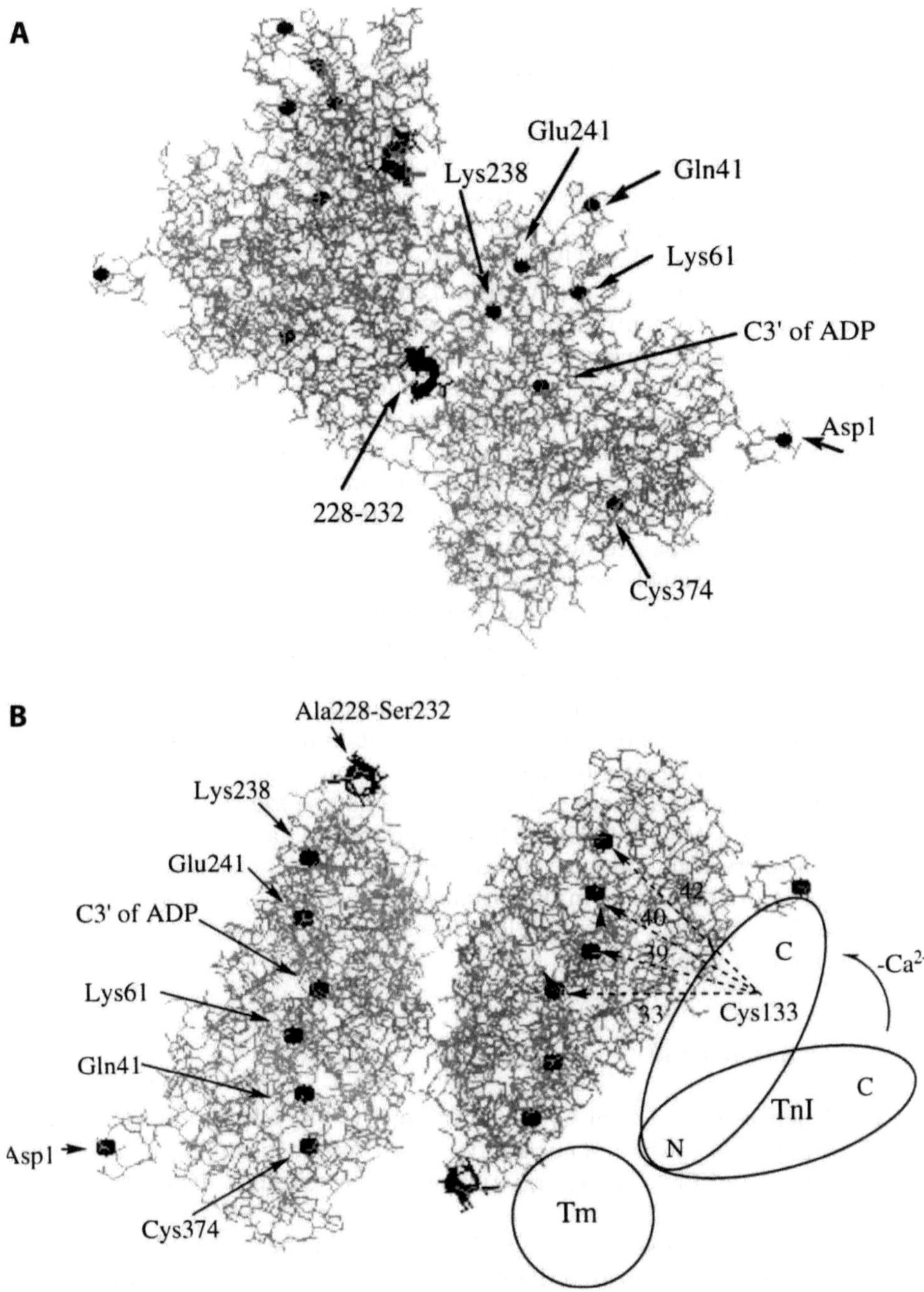

Fig. 2A,B. Atomic model of F-actin showing locations of residues related to the present work. **A** View along the helix axis, **B** cross section viewed from the pointed end. F-actin is constructed of two monomers using the atomic coordinates of Lorenz et al. (1993). Cα of Asp-1, Gln-41, Lys-61, Lys-238, Glu-241, and Cys-374, and C3′ of bound ADP are shown with *filled circles*, and residues 228–232 are shown with a ribbon structure on the wire-frame model of F-actin. The software used for the figure was RASMOL (Sayle and Milner-White 1995). In **B**, one possible spatial relationship between TnI and actin in the presence and absence of Ca^{2+} is illustrated. The position of tropomyosin (in the presence of Ca^{2+}) is quoted from Saeki et al. (1996). This figure is from Miki et al. (1998a)

the concentration of Ca^{2+} ions. This suggests that the region around Lys-61 undergoes a conformational change during inhibition or activation. This accords with the previous report that Lys-61 is important for the Ca^{2+}-mediated regulation of the actin-myosin interaction (Miki 1989).

The involvement of myosin cross-bridges in thin filament regulation was first suggested by Bremel and Weber (1972). They showed that, at high molar ratios of S1 to actin, the ATPase activity of acto-S1 can be higher than that in the absence of troponin-tropomyosin. Lehrer and Morris (1982) showed that at very low S1 concentrations, tropomyosin greatly inhibits acto-S1 ATPase activity even in the presence of troponin and Ca^{2+}. Then, in order to test a case where the release of inhibition may require the existence of significant amounts of S1, the effects of S1 (molar ratio of S1 to actin = 1/3) on the FRET efficiency between tropomyosin and actin were studied. No significant effect of S1 on the FRET efficiency was observed, whether Ca^{2+} ions were present or not. Thus, it seems that the activation of the thin filament by S1 binding is not necessarily accompanied by tropomyosin movement on the actin filament.

Three-dimensional reconstructions of electron micrographs of reconstituted thin filaments (Milligan et al. 1990; Ishikawa and Wakabayashi 1994) suggested that in the presence of Ca^{2+}, tropomyosin binds to the inner domain of actin. Using site-directed mutagenesis, Saeki et al. (1996) identified residues 228–232 and 241 on subdomain 4 of actin as the tropomyosin-binding sites. One possible location of tropomyosin on an actin filament is illustrated in Fig. 2B. The S1-binding sites on actin are located on subdomains 1 and 2 which compose the outer domain of F-actin in the Lorenz model (see review by dos Remedios and Moens 1995b). Thus, tropomyosin has to move to the outer domain to sterically inhibit acto-myosin interaction. This may cause a large change (more than 10 Å) in the distance between the probes on tropomyosin and actin (see Fig. 2). However, FRET measurements did not detect such a large movement of tropomyosin.

2.2 Troponin–Actin

FRET between Cys-133 or Cys-9 on TnI, or Cys-98 on TnC and sites (Gln-41, Lys-61, Cys-374, the bound nucleotide) on actin were measured in the presence and absence of Ca^{2+}. The energy transfer efficiencies and distances are summarized in Table 2.

The distance between Cys-98 of TnC and either the nucleotide-binding site or Gln-41 on actin increased slightly (~1 Å) when Ca^{2+} ions were removed from troponin. In contrast, the distances between the sites on TnI and on actin always decreased. Since these sites are dispersed on the outer domain of actin, the data suggest that TnI moves closer to the outer domain of actin when Ca^{2+} ions are removed from troponin. The data in Table 2 also indicate that Cys-133 of TnI is located closer to the outer domain of actin than Cys-9 of TnI or Cys-98 of TnC. Furthermore, the distance change of Cys-133 in TnI movement

Table 2. Distances between probes attached to troponin and actin in the reconstituted thin filament in the presence and absence of Ca^{2+}

Donor site (troponin)	Acceptor site (actin)	Efficiency (+Ca/–Ca)	R(2/3) (Å) (+Ca/–Ca)
Cys-133 of TnI	Nucleotide[a]	0.37/0.72	42.7/33.4
	Gln-41[a]	0.32/0.68	51.6/40.1
	Lys-61[b]	0.52/0.70	45/39
	Cys-374[c]	0.122/0.492	56.7/41.7
Cys-9 of TnI	Nucleotide[a]	0.23/0.37	49.3/44.0
	Gln-41[a]	0.20/0.28	59.1/55.4
Cys-98 of TnC	Nucleotide[a]	0.08/0.06	57/59
	Gln-41[a]	0.20 /0.17	55.1/56.2

[a] From Miki et al. (1998a).
[b] From Miki (1990).
[c] From Tao et al. (1990).

during inhibition is larger than that of Cys-9. In a hinge-like bending motion, the distance between a moving point and the hinge point does not change. Therefore, the distances between Cys-133 or Cys-9 on TnI and Cys-98 on TnC need not change to the same extent as the distance changes between these residues and actin. In fact, the distance between Cys-133 on TnI and Cys-98 on TnC increased by 5 Å when Ca^{2+} ions were removed from troponin (Tao et al. 1989). This is much smaller than the observed distance change between Cys-133 on TnI and Cys-374 on actin (15 Å; Tao et al. 1990).

From these data, we could derive the following mechanism for a conformational change of troponin. When Ca^{2+} ions are removed from TnC, the C-terminal and inhibitory domains of TnI detach from the N-terminal domain of TnC (Farah and Reinach 1995) and move towards the outer domain of actin, while the C-terminal domain of TnC does not move substantially. Thus, TnI acts something like a lever arm, and the C-terminal region of TnC locates at the hinge region. One possible spatial relationship between TnI and actin in the presence and absence of Ca^{2+} is illustrated in Fig. 2B. The distances between Cys-133 of TnI and Gln-41, Lys-61, Cys-374, and the bound nucleotide loci on actin in the absence of Ca^{2+} are also indicated.

3 A New Model for Regulation

3.1 FRET Data Suggest a New Regulation Model

FRET measurements suggest that TnI detaches from the outer domain of actin on the thin filament following Ca^{2+} binding to TnC. The time-scale of this movement correlates well with the time course of the conformational change of thin filament after stimulation, using time-resolved X-ray diffraction of frog muscle (Kress et al. 1986). Although they analyzed their data by assuming that tropomyosin, instead of troponin, changes position in an all-or-none fashion, FRET measurements did not detect tropomyosin movement. Therefore, we concluded that the detachment of TnI from the actin filament is an important event during the activation of muscle contraction, but that tropomyosin movement is not required for the activation. But this conclusion may induce the following problems. If tropomyosin does not move on the thin filament, how does the Ca^{2+}-binding signal propagate along the thin filament? Tropomyosin covers seven actin monomers along the long-pitch helix, but TnI interacts with only one (or two) actin monomers. How can TnI movement cooperatively switch the activation state of actin monomers covered by tropomyosin?

Grabarek and Gergely (1987) showed that TnI can be cross-linked to F-actin near its N-terminus, using a "zero-length" cross-linker, 1-ethyl-3-[3-(dimethylamino)propyl]carbodiimide (EDC). Also, using NMR, Levine et al. (1988) showed that the inhibitory peptide of TnI (residues 96–116) interacts with the N-terminal region of actin. There are 24 positively charged and 14 negatively charged amino acid residues in the inhibitory and C-terminal domains (96–178) of TnI. On the other hand, the N-terminus of rabbit skeletal muscle actin contains a string of four acidic residues (Asp-1, Glu-2, Asp-3, and Glu-4), which is crucial for the ATP-dependent actin-myosin interaction (Bertrand et al. 1989; Sutoh et al. 1991). During inhibition, the C-terminal and inhibitory domains of TnI move towards the outer domain of actin and make salt bridges between charged amino residues. Consequently, troponin cross-links the outer domain of one actin monomer with tropomyosin that covers seven actin-inner domains along the long-pitch helix. When the two neighboring TnIs along the long-pitch helix cross-link two actin monomers, they may cause a considerable distortion of the actin helix and/or a significant immobilization of internal motion of the outer domain of actin monomers which are located between two neighboring cross-linkings along the long-pitch helix. The distortion and/or immobilization of actin filaments may disturb the proper actin-myosin interaction during the ATPase cycle. A new regulation model by troponin-tropomyosin is illustrated in Fig. 3.

3.2 Evaluation of the Regulation Model

Although the steric blocking model is very simple and clearly explains the inhibitory mechanism by tropomyosin-troponin on thin filaments, there are experimental conflicts with this model. According to this model, the binding of tropomyosin to actin should be weakened by the binding of S1. However, the interaction of tropomyosin with F-actin is strengthened by the addition of myosin heads at low ionic strength, where tropomyosin binds very poorly to F-actin (Eaton 1976). Furthermore, acto-S1 ATPase is regulated by Ca^{2+} ions

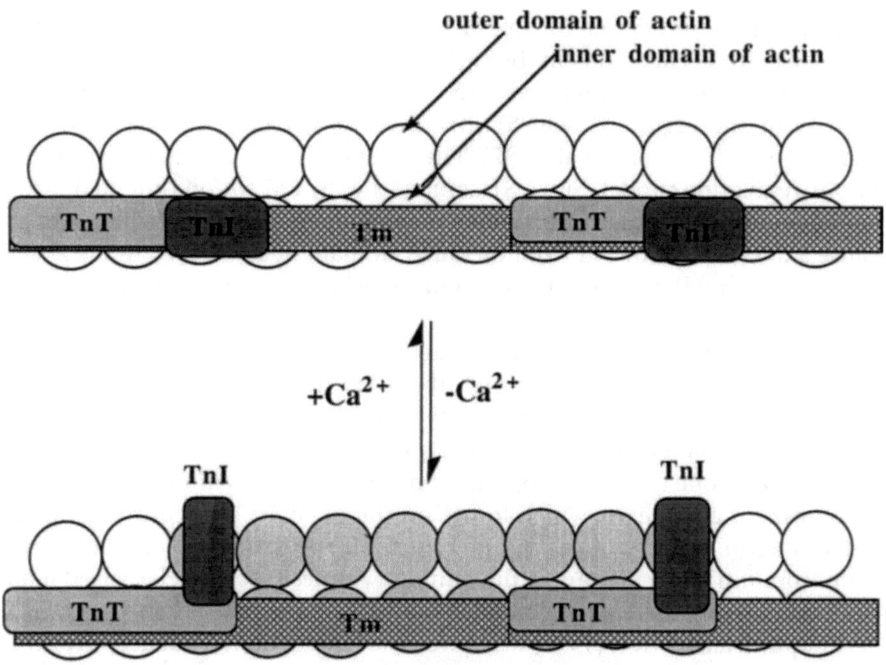

Fig. 3. A schematic model of troponin and tropomyosin along the long-pitch helix of the actin filament in the presence and absence of Ca^{2+}. One actin molecule is illustrated with *two circles* which represent the outer and inner domains of actin. The *open circles* represent the "ON" state and *shaded circles* indicate the "OFF" state of actin. One tropomyosin covers the inner domains of seven actin monomers along the long-pitch helix. When two neighboring TnI subunits along tropomyosin cross-link the outer domains of two actin monomers, the eight actin monomers covered by two TnI and tropomyosin are switched to the OFF state. This figure is from Miki et al. (1998b)

without comparable changes in actomyosin binding (Chalovich and Eisenberg 1982). S1 can be chemically cross-linked to the N-terminal region of actin by a zero-length cross-linker, EDC (Mornet et al. 1981; Sutoh 1982). The ATPase activity of this cross-linked acto-S1 can be strongly inhibited by troponin-tropomyosin in the absence of Ca^{2+} when the ratio of cross-linked actin to native actin is only a few percent. As the ratio of cross-linked actin increases to ~20%, it is poorly regulated by troponin-tropomyosin (King and Greene 1985). These experiments indicate that tropomyosin does not compete with myosin for binding to actin.

In the present model, it is TnI but not tropomyosin that competes with S1 for binding to actin. Therefore, the present model overcomes the difficulties of the steric blocking model for experimental conflicts mentioned above. On the other hand, it was reported that partial occupancy of the actin filaments with "rigor complexes" causes the release of inhibition of all actin molecules in the filament, even in the absence of Ca^{2+} (Bremel and Weber 1972). Nethylmaleimide-modified S1 (NEM-S1) has negligible ATPase activity and binds to actin permanently in the rigor state even in the presence of ATP. When the regulated actin filament was ≈50% saturated with NEM-S1, the ATPase activity of the regulated acto-S1 increased to a maximum rate even in the absence of Ca^{2+} (Greene et al. 1987). These data can also be well accommodated in our regulation model. Since S1 competes with TnI for binding to actin, the strong binding of a high proportion of myosin heads to a thin filament may cause a detachment of TnI from the outer domain of actin, even in the absence of Ca^{2+}, which follows the turning on of the whole actin monomers in the thin filament.

Ishiwata (1975) proposed a model for the regulation mechanism by tropomyosin and troponin in which the inhibition of acto-myosin ATPase in the absence of Ca^{2+} occurs through the cooperative action of two troponin-tropomyosin complexes, while activation in the presence of Ca^{2+} is brought about only by one troponin-tropomyosin complex. In his model, two troponin-tropomyosin complexes, which locate on the opposite sides of the long-pitch helix of actin filaments to each other, turn off the actin molecules only when both troponin molecules are in the inhibitory state. This model is based mainly on the following three experimental results:

1. Spudich and Watt (1971) measured the ATPase activity of acto-HMM and acto-S1, varying the ratios of troponin-tropomyosin complex to F-actin. In the presence of Ca^{2+}, native tropomyosin activates the acto-HMM ATPase activity in proportion to the concentration of troponin-tropomyosin complex. In the absence of Ca^{2+}, the extent of inhibition of the ATPase activity is dominated by essentially a two-body reaction of troponin-tropomyosin complex to F-actin.
2. Perry et al. (1972) reported that TnC removes the inhibitory action of two TnI molecules in the presence of Ca^{2+}.
3. Ebashi et al. (1972) reported that the increase of the fluorescence intensity of fluorescein mercuric acetate (FMA) bound to F-actin is proportional to the second power of the molar ratio of TnI to tropomyosin in the presence of F-actin. Although the positions of two concerted troponin molecules on the F-actin filament differs in our model (along the long-pitch helix) compared with Ishiwata's model (on opposed sides of the double helix), our model can also explain the three experimental results mentioned above.

4 Conclusions

FRET measurements did not detect any significant movement of tropomyosin on the reconstituted thin filament in response to a change in Ca^{2+} concentrations (Miki and Mihashi 1979; Tao et al. 1983; Miki et al. 1998b). On the other hand, FRET measurements suggest that during inhibition, TnI moves toward the outer domain of actin (Miki 1990; Tao et al. 1990; Miki et al. 1998a). The extent of this movement correlates well with pCa-dependence of tension development. The time scale of this movement is rapid enough to allow this movement to be directly involved in regulation of muscle contraction (Miki and Iio 1993). These data can be incorporated into a regulation model in which, during inhibition, TnI cross-links the outer domain of actin with tropomyosin which covers seven actin-inner domains along the long-pitch helix.

Acknowledgements. The author thanks many colleagues for their helpful comments and discussions including: Drs Y. Maéda, T. Kobayashi, K. Sano, T. Iio, and S. Ishiwata. The author also thanks Dr. C. dos Remedios for his helpful discussion and critical reading of the manuscript, and Dr. Lorenz for providing atomic coordinates of the F-actin filament.

References

Bertrand R, Chaussepied P, Audemard E, Kassab R (1989) Functional characterization of skeletal F-actin labeled on the N-terminal segment of residues 1–28. Eur J Biochem 181:745–754

Bremel RD, Weber A (1972) Cooperation within actin filament in vertebrate skeletal muscle. Nature 238:97–101

Chalovich J, Eisenberg E (1982) Inhibition of actomyosin ATPase activity by troponin-tropomyosin without blocking the binding of myosin to actin. J Biol Chem 257:2432–2437

dos Remedios CG, Miki M, Barden, JA (1987) Fluorescence resonance energy transfer measurements of distances in actin and myosin. A critical evaluation. J Muscle Res Cell Motil 8: 97–117

dos Remedios CG, Moens PDJ (1995a) Fluorescence resonance energy transfer spectroscopy is a reliable "ruler" for measuring structural changes in proteins. Dispelling the problem of the unknown orientation factor. J Struct Biol 115:175–185

dos Remedios CG, Moens PDJ (1995b) Actin and the actomyosin interface: a review. Biochim Biophys Acta 1228:99–124

Eaton BL (1976) Tropomyosin binding to F-actin induced by myosin heads. Science 192: 1337–1339

Ebashi S, Endo M, Ohtsuki I (1969) Control of muscle contraction. Q Rev Biophys 2:351–384

Ebashi S, Ohtsuki I, Mihashi K (1972) Regulatory proteins of muscle with special reference to troponin. Cold Spring Harbor Symp Quant Biol 37:215–223

Farah CS, Reinach F C (1995) The troponin complex and regulation of muscle contraction. FASEB J 9:755–767

Grabarek Z, Gergely J (1987) Location of the TnI binding site in the primary structure of actin. Acta Biochim Biophys Hung 22:307–316

Greene LE, Williams DL, Eisenberg E (1987) Regulation of actomyosin ATPase activity by troponin-tropomyosin: effect of the binding of the myosin subfragment 1 (S-1).ATP complex. Proc Natl Acad Sci USA 84:3102–3106

Haselgrove J (1972) X-ray evidence for a conformational change in the actin containing filaments of vertebrate striated muscle. Cold Spring Harbor Symp Quant Biol 37:341–352

Holmes KC (1995) The actomyosin interaction and its control by tropomyosin. Biophys J 68:2s–7s

Huxley HE (1972) Structural changes in the actin and myosin containing filaments during contraction. Cold Spring Harbor Symp Quant Biol 37:361–376

Ishikawa T, Wakabayashi T (1994) Calcium-induced change in three-dimensional structure of thin filaments of rabbit skeletal muscle as revealed by cryo-electron microscopy. Biochem Biophys Res Commun 203:951–958

Ishiwata S (1975) Study on muscle proteins – principally, dynamic properties of actin filament studied by quasi-elastic scattering of laser light. PhD Thesis, Nagoya University

Kabsch W, Mannherz HG, Suck D, Pai EF, Holmes KC (1990) Atomic structure of the actin: DNase I complex. Nature 347:37–44

King RT, Greene LE (1985) Regulation of the adenosinetriphosphatase activity of cross-linked actin-myosin subfragment 1 by troponin-tropomyosin. Biochemistry 24:7009–7014

Kress M, Huxley HE, Faruqi AR, Hendrix J (1986) Structural changes during activation of frog muscle studied by time-resolved X-ray diffraction. J Mol Biol 188:325–342

Lehman W, Vibert P, Uman P, Craig R (1995) Steric blocking by tropomyosin visualized in relaxed vertebrate muscle thin filaments. J Mol Biol 251:191–196

Lehrer SS, Morris EP (1982) Dual effects of tropomyosin and troponin-tropomyosin on acto-myosin subfragment 1 ATPase. J Biol Chem 257:8073–8080

Levine BA, Moir AJG, Perry SV (1988) The interaction of troponin-I with the N-terminal region of actin. Eur J Biochem 172:389–397

Lorenz M, Popp D, Holmes KC (1993) Refinement of the F-actin model against X-ray fiber diffraction data by the use of a directed mutation algorithm. J Mol Biol 234:826–836

Miki M (1989) Interaction of Lys-61 labeled actin with myosin subfragment 1 and the regulatory proteins. J Biochem 106:651–655

Miki M (1990) Resonance energy transfer between points in a reconstituted skeletal muscle thin filament: a conformational change of the thin filament in response to a change in Ca^{2+} concentration. Eur J Biochem 187:155–162

Miki M, Mihashi K (1979) Conformational change of the reconstituted thin filament – fluorescence energy transfer and fluorescence polarization measurement. Seibutsu-Butsuri 19: 135–140 (in Japanese)

Miki M, Iio T (1993) Kinetics of structural changes of reconstituted skeletal muscle thin filaments observed by fluorescence resonance energy transfer. J Biol Chem 268:7101–7106

Miki M, O'Donoghue SI, dos Remedios CG (1992) Structure of actin observed by fluorescence resonance energy transfer spectroscopy. J Muscle Res Cell Motil 13:132–145

Miki M, Kobayashi T, Kimura H, Hagiwara A, Hai H, Maéda Y (1998a) Ca^{2+}-induced distance change between points on actin and troponin in skeletal muscle thin filaments estimated by fluorescence energy transfer spectroscopy. J Biochem (Tokyo) 123:324–331

Miki M, Miura T, Sano K, Kimura H, Kondo H, Ishida H, Maéda Y (1998b) Fluorescence resonance energy transfer between points on tropomyosin and actin in skeletal muscle thin filaments: does tropomyosin move? J Biochem (Tokyo) 123:1104–1111

Milligan RA, Whittaker M, Safer D (1990) Molecular structure of F-actin and location of surface binding sites. Nature 348:217–221

Monteiro PB, Lataro C, Ferro JA, Reinach FC (1994) Functional α- tropomyosin produced in *Escherichia coli.* J Biol Chem 269:10461–10466

Mornet D, Bertrand R, Pantel P, Audemard E, Kassab R (1981) Structure of the actin-myosin interface. Nature 292:301–306

Perry SV, Cole HA, Head JF, Wilson FJ (1972) Localization and mode of action of the inhibitory protein component of the troponin complex. Cold Spring Harbor Symp Quant Biol 37:251–262

Saeki K, Sutoh K, Wakabayashi T (1996) Tropomyosin-binding site(s) on the *Dictyostelium* actin surface as identified by site-directed mutagenesis. Biochemistry 35:14465–14472

Sayle RA, Milner-White EJ (1995) RASMOL: biomolecular graphics for all. Trends Biochem Sci 20:374

Spudich JA, Watt S (1971) The regulation of rabbit skeletal muscle contraction. I. Biochemical studies of the interaction of the tropomyosin-troponin complex with actin and the proteolytic fragments of myosin. J Biol Chem 246:4866–4871

Sutoh K (1982) Identification of myosin-binding sites on the actin sequence. Biochemistry 21: 3654–3661

Sutoh K, Ando M, Sutoh K, Toyoshima YY (1991) Site-directed mutation of *Dictyostelium* actin: disruption of a negative charge cluster at the N-terminus. Proc Natl Acad Sci USA 88: 7711–7714

Tao T, Lamkin M, Lehrer SS (1983) Excitation energy transfer studies of the proximity between tropomyosin and actin in reconstituted skeletal muscle thin filaments. Biochemistry 22: 3059–3066

Tao T, Gowell E, Strasburg GM, Gergely J, Leavis PC (1989) Ca^{2+} dependence of the distance between Cys-98 of troponin C and Cys-133 of troponin I in the ternary troponin complex. Biochemistry 28:5902–5908

Tao T, Gong B-J, Leavis PC (1990) Calcium-induced movement of troponin-I relative to actin in skeletal muscle thin filaments. Science 247:1339–1341

Zhao X, Kobayashi T, Malak H, Gryczynski I, Lakowicz J, Wade R, Collins JH (1995) Calcium-induced troponin flexibility revealed by distance distribution measurements between engineered sites. J Biol Chem 270:15507–15514

Zot HG, Potter JD (1987) Structural aspects of troponin-tropomyosin regulation of skeletal muscle contraction. Annu Rev Biophys Biophys Chem 16: 535–559

Subject Index

3-D reconstructions 154
50K-20K junction 65, 68
5-iodoacetamidotetramethylrhodamine 14
64 different mRNAs 177

acceptor-labeled actin 193
actin binding cleft 23
- filaments 149, 150, 157
- mutants 33
- mutations 81
actin-myosin
- association 65, 72
- contact 65
- interface 53
- sliding 72
actomyosin
- complex 21, 53
- contacts 32
- interactions 32
- interface 28
- states 31
alanine-scanning 66
allosteric effector 123
atomic models 157
ATP
- turnover 93
ATPase active site 8
- activity 180
axial motions 8
azimuthal motions 8

blocked state 120, 123, 184

Ca^{2+}
- binding 122, 172, 173, 175
- sensitive switch 175, 176
- dependence of force 183
calcium regulation 134
caldesmon 163
calponin 163
cardiac
- muscle 52
- Tn 123
catalytic domain 8, 9
cDNAsv 181
central core disease 78
charge density 42
charged residues 34
chemical cross-linking 2
chicken gizzard RLC 11
closed state 184
conformation of actin 15
conformational diversity 100
cooperative unit 116, 122
cooperativity 112
cross-bridge cycle 46, 162
cross-linked acto-S1 199
cross-linking 121
crystal structure 176
Cys-374 24, 25

Dictyostelium myosin 65, 68
disorder-to-order transition 9, 10, 11
distortion and/or immobilization of actin 198
DNase 1 binding loop 28
DNEQ 38
donor-labeled troponin, -tropomyosin 193
dynamically disordered 16, 17

ELC 9
electron paramagnetic resonance (EPR) 8, 14, 15, 16
electrostatic interactions 53
EM 120, 121, 126
evanescent field 91
excimer fluorescence 125

familial hypertrophic cardiomyopathy (FHC) 75, 78, 134

FHC mutations 76, 77, 80, 82, 83
FHS (6-[fluorescein-59(and 60)-carboxiamido]hexanoic acid succinimidyl ester) 22
fibers 126
fibroblast Tm 125
flow-oriented actin 14
FRET
- fluorescence resonance energy transfer (FRET) 23, 98, 121, 191
- efficiencies 194
fluorescent ATP 93
force generation 72

gizzard Tm 125
global dynamics 12, 15

heart failure 77
Hill coefficient 112, 122, 123
hydrophobic interactions 66
- residues 42
hypertrophy 82
hysteretic enzymes 95

IAEDANS,5((((2-iodoacetyl)amino)ethyl)-amino)-naphtalene-1-sulfonic acid 25
in vitro motility 37, 140
- motility assay 68, 70, 134
indane-diode spin label (InVSL) 10
inhibitory peptide (96-115) 175
inner domain of actin 196
internal dynamics 12
ionic interactions 66
- strength 134, 143

kinetics 117
Km 70

laser trap 88
LC domain or lever arm 8, 9, 11, 22, 95
length dependence 139, 141, 142
lever arm 22
- model 95

maleimide spin label 10
mechanochemical cycle 55
Mercenaria RLC 11
motility 53
- of actin 37
motor
- domain 8, 54
- function 69
multiple metastable conformations 103
muscle fibers 173
mutant tropomyosin 161
myofibrils 52
myosin
- binding 31
- myosin rod co-filaments 88
- mutant 66, 69
- null cell 67
- subfragment 1 (S1 65, 114

NEM-S1 115
NH_2-terminal variants of TnT 178
NMR structure 176
nucleotide analogues 56

open state 184
orientation of myosin heads 88
outer domain of actin 197

persistence length 144
phalloidin-rhodamine 14
phosphorescence anisotropy (TPA) 8, 11, 13
photobleaching 99
power stroke 21, 53
probe microscope 90
pyrene
- excimer fluorescence 121
- fluorescence 15, 16
- iodoacetamide 14

recombinant TnT fragments 181
reconstitution 183
regulated filaments 134, 135, 139, 140, 144
regulation 150
- model 191
- of the muscle 44
regulatory
- proteins 134
- Ca^{2+}-dependent interaction 185
rigor binding 40
- complex 66
RLC 9

sarcomere 112
saturation-transfer EPR (ST-EPR) 10, 11, 13
SH1 9, 10, 24
site-directed mutagenesis 67
skeletal muscle 54
- muscle contraction 71, 172
slow conformational transition 102
smooth muscle 162
speed-pCa 134, 145
splicing 177
stereospecific bonds 60

steric blocking 152-155, 159
- theory 192
strong-binding 14, 16
sudden death in young adults 77

ternary complex 7, 12, 13, 16
thin filament 151, 154, 155, 158, 161
troponin (Tn) 134, 135, 140, 151, 152
- TnC 113, 122, 172, 173, 183
- TnI 113, 172, 173, 183
- - movement 198
- - TnC interaction 174
- TnT 113, 177, 183, 185
- - 1 178
- - 2 179
tropomyosin (Tm) 113, 134, 135, 140, 150-161
- Tm-actin 179
- movement 192

ventricular
- essential light chain (ELC) 75
- regulatory light chain (RLC) 75
visualization of single molecules 91
Vmax 70

weak-binding 12, 13, 15, 39

X-ray 120, 121

yeast actin 33
- myosin 75
- DSE mutant 38

MIX
Papier aus verantwortungsvollen Quellen
Paper from responsible sources
FSC® C105338

If you have any concerns about our products,
you can contact us on
ProductSafety@springernature.com

In case Publisher is established outside the EU,
the EU authorized representative is:
Springer Nature Customer Service Center GmbH
Europaplatz 3, 69115 Heidelberg, Germany

Printed by Libri Plureos GmbH
in Hamburg, Germany